# All About Poultry

*by Your Uncle Dudley*

**with an introduction by Jackson Chambers**

*This work contains material that was originally published in 1911.*

*This publication is within the Public Domain.*

*This edition is reprinted for educational purposes and in accordance with all applicable Federal Laws.*

Introduction Copyright 2017 by Jackson Chambers

# COVER CREDITS

**Front Cover**
*Golden-Laced Wyandotte Hen* by Ospr3yy at en.wikipedia
[FAL], via Wikimedia Commons


**Back Cover**
*Your Uncle Dudley* (Author)
Photograph from book interior.


**Research / Sources**
*Wikimedia Commons*
www.Commons.Wikimedia.org

Many thanks to all the incredible photographers, artists,
researchers, and archivists who share their great work.

**PLEASE NOTE :**
As with all reprinted books of this age that are intended to perfectly reproduce the original edition, considerable pains and effort had to be undertaken to correct fading and sometimes outright damage to existing proofs of this title. At times, this task can be quite monumental, requiring an almost total rebuilding of some pages from digital proofs of multiple copies. Despite this, imperfections still sometimes exist in the final proof and may detract slightly from the visual appearance of the text.

**DISCLAIMER :**
Due to the age of this book, some methods or practices may have been deemed unsafe or unacceptable in the interim years. In utilizing the information herein, you do so at your own risk. We republish antiquarian books without judgment or revisionism, solely for their historical and cultural importance, and for educational purposes.

# Self Reliance Books

Get more historic titles on animal and stock breeding, gardening and old fashioned skills by visiting us at:

# http://selfreliancebooks.blogspot.com/

# INTRODUCTION

It is my great pleasure to present to you another essential old volume on poultry breeding - *All About Poultry*. It was written by Your Uncle Dudley, and first published in 1911, making it over a century old now.

The book contains chapters covering topics including *A History of Some of the Breeds, Diseases, Feeding Ducks, English Breeds, How to Feed Little Chickens, Guinea Fowl, American Breeds, Raising Chickens in the Back Yard, Management of Geese*, and more.

Even although *All About Poultry* is now more than one-hundred-years old, much of the information is still as relevant today as it was when Dudley wrote it. This book contains good, solid, common sense breeding and management information from an expert on the subject.

A wonderful starter book for all those embarking on a small or large-scale breeding operation, or for those just exploring the possibility.

Jackson Chambers

*State of Jefferson, November 2017*

Your Uncle Dudley.

## PREFACE TO REVISED EDITION.

The entire first edition of "All About Poultry" was sold in less than a year, and still there is a large demand for it. I determined therefore to revise and enlarge a second edition.

In this second edition, as in the first, will be found articles that are taken from my experience in breeding poultry which reaches back over 40 or more years. Many of what I thought to be the least important articles were cut out, and in their place will be found articles more helpful and instructive to those engaged in breeding poultry in the South or who contemplate entering this profitable and attractive field.

I have sought to write so that even a child can understand clearly what is contained herein. I have avoided the use of technical terms; and in the remedies suggested for the cure of prevention of diseases, have advised the use of simple remedies that are usually found in every household and which can be administered at once without calling upon a druggist to supply them.

Thanking the people everywhere for the kindly reception given the first edition, I present this revised edition with the hope that it will be more helpful to them than was the first.

Your Uncle Dudley.

# CONTENTS.

| | PAGES. |
|---|---|
| A History of Some of the Breeds | 50 |
| A Word of Caution in Regard to Northern Poultry Papers | 15 |
| Capons | 110 |
| Crossing Breeds to Improve Them | 17 |
| Difference Between Northern and Southern Farmer | 9 |
| Diseases | 31, 117-121 |
| Ducks | 95 |
| Ducks, Indian Runner | 101 |
| English Breeds | |
|     Orpingtons | 79 |
|         The Dorking | 82 |
|         The Hamburgs | 84 |
| Feeding Ducks | 103 |
| Farmers' Wives and Cold Storage | 40 |
| Geese | 106 |
| Geese, Management of | 108 |
| Guinea Fowl | 115 |
| How to Feed Little Chickens | 32 |
| Indigestion in Chickens | 152 |
| Let Beginners Go Slow | 144 |
| Market Chickens | 150 |
| Mediterraneans | 74 |
| Poultrymen | 13 |
| Poultry Interest in the South | 14 |
| Questions and Answers | 124 |
| Raising Chickens in the Back Yard | 28 |
| Raising Little Chickens on the Cold Storage Plan | 37 |
| Reply to About 8 Letters of Inquiry | 157 |
| Scrubs and Pure-breds | 145 |
| Shape in Poultry | 46 |

## CONTENTS—Continued.

Some Questions to Answer ................................. 159
Standard Weights of Fowls ................................. 49
Strain ..................................................... 44
The Asiatics
    The Shanghai Fowl ..................................... 52
    The Brahma ............................................ 56
    The Cochin ............................................ 57
    The Langshan .......................................... 59
The American Breeds
    The Plymouth Rock ..................................... 61
    The Rhode Island Red .................................. 63
    Wyandottes ............................................ 65
    American Dominique and Buckeye ........................ 68
    The Java .............................................. 71
    The Brown Leghorn ..................................... 76
The Cornish Indian Fowl ................................... 83
The Game Fowl ............................................. 92
The Polish Fowl ........................................... 90
The Egg Question in Georgia ............................... 20
The Egg Subject ........................................... 21
The Thoroughbred and Market Eggs .......................... 24
The Poultry Product of Georgia ............................ 26
The Management of Incubators .............................. 35
To the Boy of the Farm .................................... 122
To the Wives of the Farmers ............................... 39
Turkeys ................................................... 112
Value of Influence of Pure-bred Stock Upon the Intellectual
    Life of the Family .................................... 11
What is a Utility Chicken ................................. 18

## THE DIFFERENCE BETWEEN THE NORTHERN AND SOUTHERN FARMER.

In the North and Northwest you will find every farmer giving close attention to the small things of the farm; vegetables, small fruits, pigs, chickens and, of course, eggs. He reads the agricultural and poultry papers. He not only is careful as to the wheat and corn, but endeavors to make everything on the farm count, everything points to the dollar. In the South with better soil and climate the Southern farmer can not see the small things, because he has a bale of cotton so near his eyes that it shuts out everything else. He seldom reads the poultry or farm journals. In the spring he plants about an acre of highly manured land in a garden and makes enough vegetables to feed a neighborhood. His wife attends to the poultry and if he would give her just a half chance by helping her, she would make the poultry pay. In the summer he starts to town; he fills the back and front of the buggy with vegetables, while his wife goes to the barn and in the fence corners and hunts up two or three dozen eggs, and catches half a dozen nice fries and off he goes. On the outskirts of "town" a chicken flutters, he is hailed from a house; his eggs and chicks are sold before he reaches the center of the town. He sells about one-fourth of the vegetables, because the market is glutted. Every farmer in reach of town has done the same thing. He gives them away rather than take them home or takes them back home and feeds old Brindle on them. It cost ten times as much to produce them as it did to produce the grass growing all over the fields, that could have been easily harvested for old Brindle. Now had he planted one-tenth of the acre in a garden and had made a chicken yard for his wife of the other, bought a lot of well-bred chickens—the improved sort—that would enable his wife to raise broilers and fries in two or three months, while the well-cared-for hens would produce two or three times as many eggs as do the common fowls, he would not have had any vegetables to give away in town or to feed old Brindle on such costly food. Just after the war two gentlemen were going down the Mississippi river on a steamboat. One of them was an old planter, the other a merchant, who was trying to get the old planter to plant, instead of cotton, cabbages—4,900 to the acre at five cents each, $240 per acre, with less expense than cotton. The old planter straightened himself up and with a look of intense disdain on his face, said: "Well, who in the devil would have $240 made on cabbages?" There is some of that spirit still in the South. Look around the corner of that bale of cotton, Bud. The chicken yard will not cause you to make a boll of cotton less and then the chicken and eggs from that three-fourths acre will pay you better than any two acres of cotton on the farm. So thinks your "Uncle Dudley."

# VALUE OF INFLUENCE OF PURE-BRED STOCK UPON THE INTELLECTUAL LIFE OF THE FAMILY.*

Quite apart from the enhanced market value, pure-bred stock has another value which is not always estimated at its true worth—the value of its influence upon the intellectual life of the family. One only needs to go into the family home on the farm where the pure-bred cattle, horses, sheep or swine are reared to be convinced of the reality and beneficence of this influence. If other proof be needed it may be had by comparing or contrasting a home on such a farm with one on the farm devoted to grain farming. It has been said that wheat farming debauches the mentality of the farmer. While this is probably too strong a characterization, yet it graphically suggests the mental vigor promoted by the life on the stock farm. There are some phases of the profession of farming not always well defined in our own thoughts; it is well worth while to consider some of these in their relation to the intellectual life of the farm family. For example, the circle of acquaintance and the associations which inevitably follow one's identification with any particular pure breed of live stock will widen the mental horizon; also the range of reading—imperative if one would keep abreast of the advance being made by all the pure breeds—will itself strengthen the understanding and broaden the general intelligence. Then, too, the study of nature's methods, the mysteries of heredity, the influence of environment,

---

*The above article is an extract from an address delivered by Mrs. Virginia C. Merritt, of the University of Minnesota.

bring one into intimate sympathetic touch with the great forces or laws that wait upon and reward our intelligence or perchance punish our ignorance. The more than human response in affection and absolute trust which the horse, and even the Southdown, will make to the master's care, teaches the highest lesson concerning our obligation to others. And all these lessons are so easily, so imperceptibly, transferred to other planes of life, where they influence conduct and destiny. When one appreciates intelligently and sympathetically the high privilege of controlling the conditions that create vegetable and animal life he may get a glimpse of that perfect love and perfect justice Divinity exercises toward its creatues.

BACK YARD OF WILLOW BROOK FARM PRESENTS A NEAT APPEARANCE.
[Note boards at bottom of division fences.]

## POULTRYMEN.

It is not an uncommon thing to hear complaints from those who have just commenced to breed poultry as to having trouble with chickens and eggs received from those who have been in the business for a number of years. Sometimes harsh things are said and from the want of a complete knowledge of the facts of the case charges are lodged against a shipper who has done his best to do the right thing and who is conducting an honest, square business. For about thirty-five or forty years I have been brought in close contact with poultrymen all over the United States. In the South I am personally acquainted with a very large number of the most prominent and also of the lesser lights among those engaged in breeding fancy poultry. I am quite sure that those engaged in this business are above the average in honesty and fair dealing of those engaged in any other business carried on in this country. That there are perhaps a small number of stricksters in this business there is no doubt whatever. But they can not possibly remain in business for any length of time. Now, the poultry business is not conducted like any other business that I can now call to mind. Every man in it is an independent dealer—no trust or combine, or "special interest" here. And yet there is among these men a settled determination, in order to protect themselves, to expose every man who acts at all shady in the transaction of his business. Again, poultry journals everywhere are conducted by men who necessarily have been engaged in breeding poultry. They must thoroughly understand the business in its every detail, and they are compelled to protect every man who has an ad. in his paper and also every subscriber. The moment that a dishonest chap turns up and the fact of his dishonesty is established, it at once becomes absolutely impossible for him to get an ad. in any poultry journal anywhere. Recently, a man appeared, advertising largely in many poultry papers, claiming that he had the best Rhode Island Reds on earth and that he had a large number of blue ribbons to his credit. He shipped chickens and eggs all over the country. He shortly reached his limit. *The Southern Poultryman* landed on him on a complaint of unfair dealing from one of its subscribers. That paper promptly returned him the money paid on a contract and discontinued his advertisement. There is not to-day a poultry journal anywhere that will handle his ad. business. So you see, that while a very, very large percentage of poultrymen are honest and stand for that which is clean in business, the few shysters are compelled to act honestly or quit business.

When, therefore, you give an order for chickens or eggs and there is anything wrong, write the shipper a nice, kindly worded letter of complaint, and ninety-nine times out of one hundred you will be made happy by a quick adjustment of your trouble.

## THE POULTRY INTEREST IN THE SOUTH.

In the tier of States known as the Cotton States recently there has been an astonishing awakening in regard to breeding fine poultry. For many years the interest in well-bred chickens has lagged. Poultry journals were born only to die for want of advertisements and subscribers.

Chicken shows attracted so little attention that it did not pay to hold them—for almost invariably only "the cranks" attended them and they invariably had to foot the bills, except where they were held in connection with

State or county agricultural fairs. And even then poultry was not even reckoned among agricultural products.

Farmers and many of the town folk, ridiculed the fellow who "fooled away his time" on "fancy chicks." Looking backward from the standpoint of to-day it is simply marvelous that such a state of affairs could have existed in a country with the advantages that these Cotton States possess.

The state of affairs as mentioned above changed slowly for the better up to two or three years ago, when a real revolution was inaugurated. It rapidly increased in power and intensity. The old farmer visits the poultry shows to-day, adjusts his glasses and looks into a chicken coop. "How much for that rooster?" "Five dollars," answers the owner. "I'll take him, for he is just the chicken that I want." Who ever heard of such a thing four or five years ago?

Well, the shows have won out, and the fancy are coming into their own.

The "fairs" used to place the chicken coops in the rear, where no one could find them. To-day they are fully recognized as an agricultural product, and are placed in a conspicuous place at all of the fairs.

## A WORD OF CAUTION IN REGARD TO NORTHERN POULTRY PAPERS.

Not for one moment would I say a word against Northern poultry journals as being adapted to the needs of Northern poultrymen and others engaged in the work who are breeding poultry for profit or pleasure. There are a very large number of them published all over the North. Some of them are published simply as a scheme to advertise a new system, or an incubator combination, or perhaps a combination that controls several poultry appliances. Some of them, so far as I have been able to find out, are published strictly in the interests of all those who are engaged in raising poultry. Now group them all together and what do you find? Poultry journals, elegant in outward appearance and filled from the first page to the last with sensible, practicable, well-written articles, that in a large measure apply exclusively to the North. Of course, there are some of these articles that are helpful to the Southern poultry breeder, and to one who has been in the business for some years and who can, from his past experience with them, use those that are helpful to him in the South, and discard those that apply only to the North. But how about the beginner? He is just starting out, his ears and eyes are open. He does not know how to do the simplest and most ordinary things. He subscribes to several Northern poultry journals. Difficulty after difficulty confronts him. He reads a Northern journal, and gets hold of that which puts him into greater difficulties. He becomes discouraged and quits in disgust. I am giving largely my own experience in the above. I frequently ask myself the question, "How long before an end will be put to the miserable habit of going to the North for everything that we need?" This habit seems to have fixed itself upon us all. These Northern journals have thousands of subscribers in the South, while our Southern journals "languish and pine" because Southern people do not support them. Those who have crossed over the line are now looking backward and are singing lustily "never again." They are trying to warn the beginner not to do as they did, but to patronize Southern monthly journals, which treat exclusively on Southern conditions.

## CROSSING BREEDS TO IMPROVE THEM.

People everywhere over the South, and perhaps elsewhere, have some strange ideas concerning crossing the different breeds of poultry to improve them. You can hear persons who perhaps have been raising chickens for a long time say: "I have a better breed of fowls in my yard than any that I know of," and then they will go into the details and tell you how they crossed one breed with another and then describe the product of the cross. One who has a perfect knowledge of the breeds that he has put together sees in a moment that he has fowls that are inferior to either of the breeds that he has used. These folks do not for a moment consider that it took years of careful and skillful crossing, with a definite object in view, that is, to produce a fowl that could be depended upon to do certain things and to be superior to other breeds on the same line. He does not stop to think that by crossing two of these nearly perfect breeds that he is very near where the originators commenced to improve them. In the product the result of this boasted cross is that he has a chicken that is no better than the original stock that the skillful and patient breeder commenced with, to produce a fowl that was better than any fowl in its class. In other words, they are not much better than the common barnyard fowls that he could have bought for fifty cents. Again, I have stood by and seen farmers and others pay good round prices for a large Rhode Island Red—or of some other breed—male, weighing eight to ten pounds to cross with his small common hens, with the idea of breeding up to a lot of fowls that would equal in every respect their sire. There is no question that he improves his flock in the first generatio, and if he uses a new male the next season he will continue to improve them. But what a slow and tedious process! About one-third of the result will be like their mother and by a slow process he will have a fowl superior to the barnyard fowl, but far inferior to the sire that he commenced with. Now, for a few dollars more he could have purchased a trio and saved a world of trouble trying to produce an impossibility, and that would have made him, with the pure breed, many times more than the price he paid for the trio. It is wise, therefore, to pay a little extra for the thoroughbred fowl, that the other fellow has the trouble and expense of producing.

## WHAT IS A UTILITY CHICKEN?

There is, to those not informed as to the terms used by poultrymen, a great misconception as to what is really meant by a utility chicken. Every one knows what "utility" means, and catches the idea at once, when advertised by poultrymen—that it means a general purpose fowl. But now how is that fowl bred? Is it a scrub? Is it bred apart from the show birds and are the show birds any better layers? Do the show bird's chicks grow faster? Well, let's see if I can explain what a utility bird really is, and how it is bred. In every hatch—large or small—from the oldest strain and the best birds used in America there will be chickens with some slight defect; sometimes it happens that a bird is as fine specimen as there is in the poultry yard, but there is a white feather or several of them where there should have been red feathers. Then, again, perfect except a feather on the leg, or a twist, or a lump on the comb, or a hundred and one small, unimportant imperfections that make the bird unfit for show purposes, but it is from the same parents, and full brother or sister to the bird that goes out of the show with a blue ribbon, and thereby enhances his or her value many times.

These birds will not do to breed from by a poultryman, because he is after, as nearly as possible, mating up his birds so as to produce birds that will conform to the rules laid down in the "American Standard of Perfection," in which every bird is described so perfectly that no one has ever attained to it. Frequently a breeder uses these slightly defective birds with splendid results. For instance, he, in making up his breeding pens, finds a hen from a long line of almost perfect parentage that has a defect in her comb. Well, he has a cock that has a perfect comb, and so where one is weak and the other is strong, he often gets the best results from such a mating. Sometimes it happens that utility birds produce equally as perfect birds as those of the same blood, mated for show purposes, but they are not apt to do this, and the breeder can not afford to take any chances, but when it comes to egg production or breeding fowls for market they are equal to any blue ribbon chicken of the same strain.

Good Utility Birds.

## THE EGG QUESTION IN GEORGIA.

It is a matter of surprise to one not informed when he goes into the statistics as given by the agricultural department of Georgia as to the egg product of the State, to find such a great difference in the product of the different counties. The latest report is that of 1899. There have been some changes, of course, and within the last two or three years a decided improvement, but the improvement is, from the best information obtainable, not as great as it should have been, and the relative situation remains about the same.

A rather remarkable feature in this report is this: That the counties in which are located the large cities produce fewer eggs than counties of the same size or smaller in other parts of the State. For instance, Fulton produced only 122,460 dozen eggs; Bibb, 82,090 dozen; Chatham, 65,570 dozen; Richmond, 84,140 dozen; while Cobb produces 266,710 dozen; Houston, 118,940 dozen. The small county of Bryan, adjoining Chatham, 71,890 dozen; Burke adjoining Richmond, 278,330 dozen. Then Glynn, 13,740, and Wayne, 67,040. You have, then, Atlanta, Macon, Augusta and Brunswick each in counties that produce far fewer eggs than the surrounding counties.

The farmers just at their best markets are allowing the adjacent counties to go over them and supply markets that are first at hand. Now this is said in no spirit of criticism of the counties adjoining the counties with the large cities in them, because all of these cities use thousands of cases of eggs shipped in from other States. These facts are only given to inform the "near-town" farmer what he loses by not producing more eggs.

Carroll county produces 421,230 dozen eggs; Glynn, the smallest number, 13,740 dozen. The next largest is Gwinnett, with 335,880 dozen. The first ten counties produce over 200,000; there are fifty-seven that produce over 100,000, and sixty-eight that produce less than 100,000 dozen. The total production of the State was 15,505,303 dozen.

Now, the above figures are taken from the last report of the agricultural department, and no doubt the next report will show a great improvement over the report from which these figures are taken, but the fact remains that the farmers of Georgia are far behind other Southern States in regard to the very important matter of breeding poultry.

## THE EGGS SUBJECT.

The country was startled and perhaps shocked when it saw in the newspapers that thirty-six million eggs and several tons of poultry had been discovered in a cold storage plant near New York City that had been on storage for over a year. To folks who know what this means and are posted as to cold storage eggs and poultry, a fearful menace to human life is seen in this statement. An egg taken from cold storage will decay very quickly, when it becomes absolutely useless as an article of food. The people all over the country are rapidly becoming acquainted with the facts concerning cold storage eggs, and therefore the extension of this knowledge is increasing the demand for the fresh article. Should any one wish to prove the above statement, let him try to poach an egg just out of cold storage; it can not be done. Some of the hotels and restaurants use these eggs to scramble. Two days out of storage and they will not do to fry. They are then used for cakes and served in the scrambled form.

The above facts, if known and acted upon by the farmers of the South, would undoubtedly cause them to produce more eggs and chickens, and make them a part of the main crop. As the case stands to-day, no one ever heard of a farmer who carried eggs back home because he could not sell them; but properly handled, he could increase the price of every egg that his poultry can produce to ten cents above the market price and have regular customers. The date when laid is marked on every egg. They are nicely packed in a paper carton that holds just one dozen, and the demand will always far exceed the supply. Other poultrymen are doing the same thing and their delighted customers say they are always glad to have them. The price now is about fifty cents per dozen. I mention the above because I know the above statement to be a fact.

King Cotton stands alone with the Southern farmer. His majesty needs help in order that he may properly carry on the government. He would like to create several new peers—the Earl of Pigey and the Duke of Roosteroso, and Lady Hen, and her daughter, Lady Pullet, and Sir Capon. Bringing these royal assistants to his help, King Cotton would then be such a mighty power that the meat trust and cold storage would be relegated to the rear.

There is ot-day a very serious question confronting the Southern farmer, that they should at once look into and correct. Recently, there has been all over the country, North, South, East and West, a determined effort on the part of the United States Government, and then by the governments of the different states and also by many prominent individuals, to give the people of the entire country pure food. Georgia is making a splendid fight against impure grain for the farmer, and impure food for everybody. Every citizen of Georgia should do everything in his power to assist in this great fight. Pure food means much to each individual, and as the farmer is the producer of everything that we eat, it is but reasonable that we look to him to get into the front ranks and do valiant service, not only for his own, but also for his neighbors' protection and so, I desire to call his attention to some facts about eggs. Now the question that should deeply interest him is, what sort of eggs am I sending to market? And what sort of eggs are my wife and children eating day by day? So in this article I purpose to give facts that will probably startle some folks, but you can not disprove a single fact that I give. Eggs are porous, and they absorb bad odors; if laid in a filthy cowpen they absorb part of the filth. If the nest is out among the noxious weeds they absorb the noxious odors. Put one or two drops of turpentine on an egg and you can not eat it. Kerosene oil will produce the same effect. Did you ever milk a cow in a filthy cow stable? Well, if you never have done so, I want to tell you that you can detect the foul odor in the milk as soon as you get to the house, and yet eggs take in bad odors more quickly than does milk. Again, as to feed. Every farmer knows that in the spring, if the cows eat bitter weed or wild onions, what the result will be to the milk. Why? Because everything that the cow eats goes into the milk. Now why not everything that a hen eats go into the eggs? Well it does. Laying hens kept about a filthy lot, taking into their crops and breathing into their lungs this filthy air, necessarily become saturated with filthiness. Then how can their eggs be pure enough to eat? Are we all really bent on having pure food? Well, says the farmer, how can I remedy this? Have a yard built for your laying hens, a house for them to roost in, feed them on grain and green food, make nests around the fence for them, do not let them lay in the hen house, empty the ashes in the yard, spade it up once a week. When you go to town tell the folks what kind of eggs you have for sale. Go to every doctor in town and tell him that you have eggs, packed in boxes, one dozen each, with the date that the egg was laid marked on the egg; that the hens were fed on pure, clean food, that you guarantee them to be fresh and clean inside and outside. He will want them for sick folks and you will soon build up a trade that will enable you to get for every egg taken to market, ten cents a dozen advance on the market price.

---

## THE THOROUGHBRED AND MARKET EGGS.

By request, Mr. E. W. Burke, of Macon, Ga., breeder of Buff Orpingtons and Rhode Island Reds, wrote the following:

"Oh, I don't want any fancy chickens; just want some to lay eggs for the family; can't afford to go in for the fancy." Ever hear anybody say that? Let's see which is cheaper, which pays best. If a hen can be maintained at the cost of $1 per year, she will pay a profit, if she lays eight dozen eggs in a year; if she lays ten dozen, she will pay fifty per cent. more profit. During 1909 the price of market eggs did not go below twenty cents, and the average price was about twenty-six cents, so that the eight-dozen-egg hen earned above her feed $1.08; the ten-dozen-hen, $1.60, and the twelve-dozen hen, $2.12. The fact is that the twelve-dozen hen earned more than $2.12, because in order to produce 144 eggs she had to lay some of them during the cold months, when eggs were higher. The cheap hen, costing about $1 and $1 to feed, will only go to eight dozen in a year; most frequently not over six dozen, and the majority of them lay during the pleasant months, when eggs sell at the lowest prices. Again, eggs from mongrel fowls are not uniform in size or color and do not command the price of extra fancy eggs.

The figures used in the foregoing estimate are based on eggs at wholesale market, or case egg prices; but there is a better profit than this to the egg farm near a city, for special customers can be had to take guaranteed eggs gathered daily and sold before they are twenty-four hours old, and they will easily pay from five to ten cents more a dozen for them above the retail market price. In Macon the past few months, I have sold guaranteed eggs at forty to fifty cents a dozen.

To command these good prices, eggs must be gathered daily, packed in cartons of one dozen; each carton sealed and every egg under the seal guaranteed to have been laid on the days marked on the carton.

Doctors are prescribing egg diet in so many cases that the demand for strictly fresh eggs is greater that it has ever been, and it is steadily increasing.

Pure food is the order of the day, and an egg should be as free from impurities as any other article of diet. An egg farm must have yarded, grain-fed birds, and not the barnyard scavenger that produces the usual "fresh eggs" on the farm. Germs can as well be transmitted by the egg we eat as by the milk we drink.

The doctors are appreciating this fact and the invalid in only too glad to be able to get the pure and palatable egg at an advance in price over the ordinary kind.

E. W. BURKE,
The Oaks Farm, Macon, Ga.

## THE POULTRY PRODUCT OF GEORGIA.

There are some facts concerning poultry breeding that should be known to every farmer in the South. I believe that it is the experience of poultrymen that all things considered, the production of eggs pays better than breeding chickens, and yet, with intelligent management, broilers and fries sent to market early in the spring pay a handsome profit. Later on, when the market becomes well supplied, eggs take the lead, so far as the profits are concerned. Your early hatched pullets will make your winter laying hens, and if you can hatch off your chicks in December, January and February, they at eight or ten weeks old will bring from fifty to sixty cents each. It takes close attention and careful management to accomplish the above results, but there are only a few men in the South doing this to-day, and if one can do it, all can, if the right sort of effort is made. Now, in a previous article, I gave the latest reports published by the agricultural department of Georgia as to the egg crop of the State, and from the same report I find that the last report on poultry is as follows: Fowls over three months old, bred in one year, 4,549,144; turkeys, 103,416; geese, 208,997; ducks, 64,895; and again the same thing obtains largely as in the report of the egg production. The counties in which are located the cities, produce fewer chickens than the surrounding counties. Fulton, 39,120 fowls; DeKalb, 45,375; Bibb, 21,363; Monroe, 40,425. This is a large county and so is Houston, with 48,655, but every county touching Bibb produces more poultry than Bibb. Richmond is an exception, for it is bordered by two large counties and three smaller counties, two of which it exceeds, and one of which—Hancock—more than doubles Richmond.

The Glynn product is only 5,668 (this is not a good county for Methodist preachers), while the small county of McIntosh goes to 8,342. Again we find Carroll leading every county in the State in poultry with 108, 292, and Glynn with 5,688.

Now from the above figures, taken, as I have stated, from the latest and best reports that we have as to the egg and chicken product of Georgia, we easily see what a splendid business opening is presented to the farmers, for all the counties are producing one-tenth or less, perhaps, of chickens and eggs than the people of the State consume; a product, too, that is as saleable as cotton, and that is always sold like cotton, for spot cash.

CHICKENS ON THE TOWN LOT.

## RAISING CHICKENS IN THE BACK YARD OF A TOWN LOT.

Some forty years or more ago, I commenced as a pastime to breed poultry on a small town lot. I had to rely upon my friends for advice, almost every one of whom had his own way of doing things, very few of them agreeing on any one point. I was, therefore, largely thrown on my own resources, and so what I am going to say about raising chickens in a back yard is the result of my own experience. I wish to state that for several years past I have not lost a chicken that came from the egg in good condition, except from my neglect, and that seldom occurred.

There are several things that are absolutely essential in the matter of raising chickens when kept closely confined within the dimensions of a small back yard. First, you must be in love with this delightful pastime—so much in love with it that you will not in the least degree think that anything connected with the work is troublesome, but on the other hand, is, to a tired business man, recreation—pleasure. Second, never forget that your chickens are in close confinement, leading artificial lives, and that they are entirely dependent upon you for such things as are essential to their health and general well being that they could easily obtain on a larger range. In the main they need lime, supplied by cracked oyster shells, which will supply them also largely with grit; then, as a back yard is always hard and will necessarily soon become foul, it will need frequent spading. Then crushed bone and an abundant supply all the year 'round of green food; fresh ground bone about once or twice a week or dried beef scraps, if the fresh article can not be had. Fresh ground bone fed every day would not injure them, but if suddenly discontinued will cause them to eat their eggs. An abundant supply of clean, fresh water is essential. They should be supplied with crushed charcoal, which, mixed with the oyster shell, will please you to see how readily they eat both. Now, if you are interested in chicken breeding in your back yard, cut this out and keep it for it will be followed by other articles, telling you how to set the hens and care for the chicks.

In a former article I told you how to prepare the yard and also some of the things that were essential in order to make a success in the above pleasant and profitable pastime. Now as to the best breed of fowls to use: My advice is that fowls like the Rhode Island Reds, Orpingtons, Wyandottes, Plymouth Rocks and Langshans are far superior to any others. They stand confinement better than Leghorns or any of the non-setters. They are quiet, very gentle in disposition and are easily managed; and then they afford you so much pleasure in raising the chickens. There is no great difference in the general characteristics of the above varieties named. Having selected the variety, purchase six or eight hens and a rooster from a reliable dealer, and with a hen-house built as nearly as possible according to the following plan you are ready for business. The above number of fowls will give better results than a larger number. It will insure you a large percentage of fertile eggs, and prevent crowding, which causes disease. For the above number you will need a house, say, six by eight or ten feet and eight feet high. It should be built of good seasoned lumber. Two sides should be entirely closed; avoid having cracks in the solid sides; a small draught will give fowls a cold that will end, unless checked, in roup. The front and west side should be slatted—the slats two inches apart, or wired with two-inch mesh poultry wire. In winter the open slides can be covered with a curtain of cloth. The roosts should never be more than three feet from the ground. Build a broad shelf two feet high in the back of this house and on it erect your roosts—have the roost on a movable frame, like a carpenters bench, cover the shelf or drop board with sand —remove the droppings every day. There should be no nests in the hen-house. Be sure to have the roof tight. The roof should slope about one and a half feet.

Now about the treatment of poultry confined in the back yard. Hens like to be in a quiet place at all times. They soon become wild and hard to manage if they are disturbed. Kindness and gentleness soon teach them that you are their freind, and in a short while you can pick them up and caress them, or minister to their needs in sickness. Never lose sight of the fact that they are utterly dependent on you and that they highly appreciate any kindness that you show them. There should be shade in the yard or some place where they can get out of sight for one or two hours at noon.

Avoid soft food entirely if possible. In the morning, as early as possible, give them the mixed commercial feed but never more than they can eat up clean. At noon feed as in the morning. At evening whole corn, oats or other grain. Be sure that they get enough and leave none on the ground. In the winter, for green food, soak oats in water until they sprout and commence to show the green, and they will eat roots and all. Alternate with dry alfalfa meal dampened, mixed with wheat bran and oats as a midway feed. Do not forget to keep crushed charcoal and oyster shells, cracked, always accessible to them. Keep also in a box three or four inches deep charcoal, fine, one inch deep, one-half inch sulphur, with wheat bran sprinkled over the sulphur. Keep this out of the rain, where they can get it. A frame to fit just inside the box, covered with inch and a half poultry wire, will prevent them from scratching the contents out. Now, of course, no one can give explicit directions as to feeding day by day, and the above may be varied according to circumstances. However, small grain, wheat, rye, oats, etc., may be fed separately. Oats is about the best of them all. Ground bone, fed with wheat bran, dampened, three or four times a week. Fresh lean meat, liver or dried meat scraps two or three times a week. Keep the yard spaded up. Feed on the loose ground, and give them fresh clean water all the time. All of the above is intended only as an outline to act as a guide for the man who keeps chickens in his back yard of a town lot.

## DISEASES.

The disease that gives great trouble in raising chickens is roup. It begins with a cold, and if not checked will spread rapidly until the entire flock is affected. In its first stages it is easily cured, later on it is almost impossible to cure. Almost every poultry man has a different remedy. I have been successful with the following: Sulphur, lard and pulverized charcoal, one part each, made into a pill as large as your finger and half as long, given morning and night for two or three days. Five grains of quinine with first dose. Then, with a dropper such as you use to fill your fountain pen, inject chloro naphtholeum, one part to twenty parts of water, in the nostrils, and in the slit in the roof of the mouth. The affected fowl should be confined, apart from the flock, for the disease is highly contagious. In the second stage of the disease you can easily detect its presence by an offensive odor from the mouth.

Sore head is a summer trouble, and as it generally appears suddenly, affecting, perhaps one-half of the flock at once, a close watch may enable you to arrest it at once, but if allowed to get a firm hold on the flock it is difficult to cure. It commences with a number of small warts on the head. Use the same treatment as for roup, except that you paint the warts with the chloro naphtholeum solution. Of course both of these remedies are for grown fowls. Lessen the dose, therefore, according to the size of the chicken.

Gapes can be easily cured by giving the chicken a lump of cooking soda about the size of a pea.

Sour crop is caused from indigestion. The crop becomes enlarged because the food does not pass to the gizzard. Feeding soft food is largely responsible for it. Remedy: Fill the crop with warm water and then hold the fowl's head down and gently squeeze the crop until it is empty. It sometimes requires two applications of the water to cleanse the crop thoroughly. A little cooking soda should follow the cleansing of the crop.

The above three diseases are those that give the most trouble. To cure the others, try to find "Your Uncle Dudley."

## HOW TO FEED LITTLE CHICKENS.

I am going to give in this article my way of feeding newly hatched chickens, and then in another how they should be housed on the "cold storage" plan. Feeding and housing are almost inseparable; therefore, you will have to read both articles in order to arrive at a proper understanding of this all-important part of a chicken's life. In telling how to manage an incubator, I said that the hatch largely depended upon the first seven days during incubation, and so the lives of the little fellows depend very largely upon the first ten days of their existence.

For the first twenty-four hours they subsist on the yolk of the egg, and require no other food whatever. They should be taken from the incubator or from the hen and placed in a coop, which will be described in the next article, and given only fine, natural sand or grit with wheat bran or alfalfa meal sprinkled over it. I do not like the artificial or manufactured grit. Now, the object of this is to enable them to get sand in their gizzards before any food is given them. God gave them, you see, a mill inside of them, but He did not furnish them with the rock to operate it, so if you will give them the opportunity they will take in enough grit to run the mill after twenty to thirty hours. As soon as the mill is ready for business, give them a little small grain. I have never found anything better than small cracked rice, but any small grain will do, except cracked corn or corn in any shape whatever. Never under any circumstances feed on soft or medicated feed. Pure wheat bran can be sprinkled over the sand with the cracked wheat or rice. All along through their lives they will need wheat bran. Your object should be to develop every organ at the same time; so you give them the rock for their mills and then something to grind, but when you give them wet meal their mills are almost choked up and the soft stuff will sour. Then comes indigestion and then death, but keep the mill busy with grain and it will develop, while scratching in the wheat bran for the grain to grind develops the muscles and all of the organs move along together, as nature demands that they should. Charcoal (fine), green food, fine crushed oyster shells, and, of course, fresh, clean water, should be kept before them constantly. Continue to feed in this way from ten to fifteen days, when you can feed the ordinary commercial feed. But see to it that there is very little corn in it. Dry alfalfa or sprouted oats will furnish green food. A shallow box with cracked charcoal at the bottom, sprinkled with sulphur and covered with wheat bran, should always be before them.

The little chickens are now about fifteen days old. They have been fed with small grain, green food, pure wheat bran and have had constantly before them crushed oyster shell, sand, crushed charcoal, and a supply of clear, fresh water, but no soft food, no corn in any form. There are about twenty to twenty-five in each coop. They now are getting to know you, and when you go near the coop they all run to meet you. You have laid the foundation for a good, strong, evenly developed constitution, and they feel good. Their mills are working beautifully, their muscles are becoming strong and they need a slight change of diet, and so you will have to feed them on the commercial food, composed of wheat, barley, rye, clipped oats, but see to it that the kind that you buy has no, or very little, cracked corn in it. You will find sometimes that it contains oyster shell. Crushed oyster shell is worth $1 per 100 pounds, while the grain food costs about $2.25, or over, per 100 pounds, so buy your shell apart from your grain. They will not eat it up clean for the first eight or ten days and you will have to move the coops about every other day to a place that has been spaded up, and raked off, so as to have nice, clean quarters with a fresh supply of grit. They will now require fine ground meat about twice a week—lean meat if you can get it—and if possible, ground fine with green bone; avoid giving them fat meat. Should you be unable to get the fresh meat, give them dried meat, which you can get at any time from the poultry supply stores. Never give fowls or chickens the medicated prepared foods advertised to make hens lay and chickens grow fast. Good, sound grain will do all that is claimed for this stuff, much better than medicine will.

As the chickens grow they should be separated and given larger quarters. They should never be crowded. A great many chickens are lost by confining them to close quarters. They should be given room enough to enable them to move out from the crowd when they get too hot. When they get to be six weeks old let them have as open place for their roosting place as possible, shielded from wind, drafts and dampness. You will, of course, have to depend upon your judgment for many things that can not be told here, but just think, and then use good, old-time horse-sense and you will succeed.

## THE MANAGEMENT OF INCUBATORS.

The following remarks as to managing an incubator are intended for those who are just starting in the poultry business, and not for the old-timers. The directions given are in a general way the main points, for, of course, I could not cover the minor points in an article of the length of this:

The most important thing to do is to buy a first-class machine; cheap machines are often worthless. Your machine should be operated in a place where you can, at all times, maintain an even temperature, and where it would never come in contact with a draft. It should be beyond the reach of bad odors. The room should be dry. The machine should be level and steady. It should be opened and aired before putting the eggs into it. You should be careful and under no circumstances put a soiled or an imperfect egg in it. It should be set at a temperature of 102½. Never pile the eggs one on another, but fill the trays with clean, fresh eggs. During the first seven days the eggs will require careful attention. They must be turned two or three times daily. This must be quickly and carefully done, for should they become chilled it will seriously affect the hatch. At five days the embryo commences to take life, which on the seventh day can be plainly seen. (All first-class machines send egg-testers with the machines.) Every egg should be carefully tested now. A perfectly clear egg is infertile; a dark or cloudy egg is imperfect. These should be removed. The infertile egg can be used for food and the cloudy egg should be destroyed. At the beginning of the second week the temperature should be raised to 103 and kept at that figure until the hatch, except that during the latter part of the second week and along through the third the ventilators should be opened for a short time to give the eggs a fresh supply of oxygen. On the twentieth day some of the eggs should be pipped, and on the twenty-first all should be hatched. On the twenty-first day, in the morning, examine those not pipped; hold the egg in your hand, and should you feel a slight movement in the egg, make a small hole in the butt end, being careful to use an instrument that will not penetrate far enough to touch the chick. Look in, and if alive, carefully help the little fellow to get out; but be careful not to cause the inner skin of the egg to bleed. Just open the egg enough to allow the chick to free himself. In a first-class machine you will not need to supply moisture, except just before hatch, when a clean cloth (not dyed), dampened with warm water, can be spread over the eggs.

I will say something about caring for little chickens in my next.

## HOW TO RAISE LITTLE CHICKS ON THE "COLD STORAGE" PLAN.

I have contended for years that artificial heat was not only unnecessary, but hurtful, to newly hatched chickens. I have written much on the subject, and while no one paid any attention to me, I just plodded on in my way of feeding and brooding without lamp or hot water. I seldom raise less than one hundred per cent. of the hatch that come from the egg in a healthy condition. In the article just in advance of this I have told how I feed them, and now I give you the cut of a simple coop that I have been using successfully for some years back. As you will see, the coop is lightly built. It has a door on the front, which is solid, except the door. The sides and top are covered with one inch mesh wire; on the back end is a box, the top of which is tight, but on hinges so that it can be raised, thus giving you access to both ends of the coop. The box should be two inches from the ground. The coop for twenty-five chickens should be three feet wide and five or six feet long, and about ten or twelve inches high. The

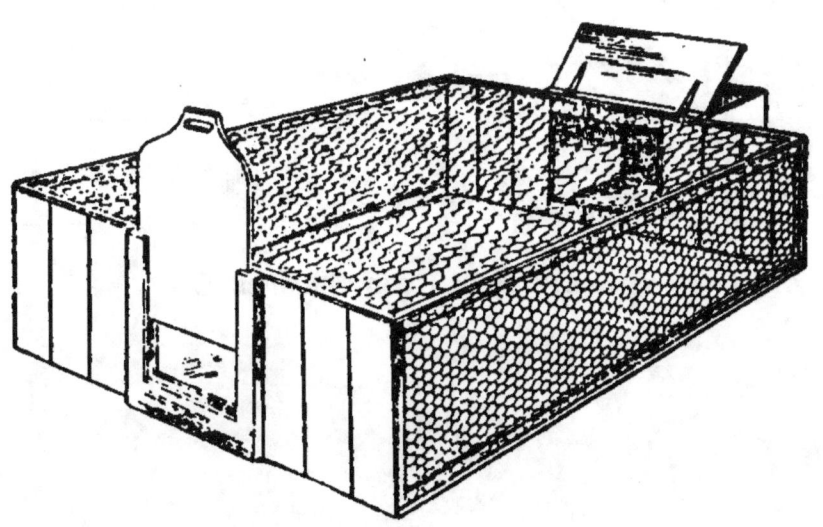

A GOOD BROODER COOP.

nest box at the back end of the coop should have an inside opening of six inches long and four inches high. There should be a small shed above the opening to keep the rain from beating into it. There should be on top of the coop a light water-proof cover—a frame covered with roofing paper will answer. This should not be nailed to the top, but loose, so that in bad weather the coop can be kept dry, and the front of the nest box also protected. The nest box for twenty-five little fellows, should be about eighteen or twenty inches square, and sixteen to eighteen inches high. A little fine excelsior on the bottom will add to their comfort. On every cold night cover the coop with burlaps. Be sure to have the top of the box to shut down tight, so that a draft will not reach them. Now, put them in as soon as twenty to twenty-five are dry, feed as in former instructions and you will raise them all if you are careful. Watch them for one or two nights, to see that they all go to bed.

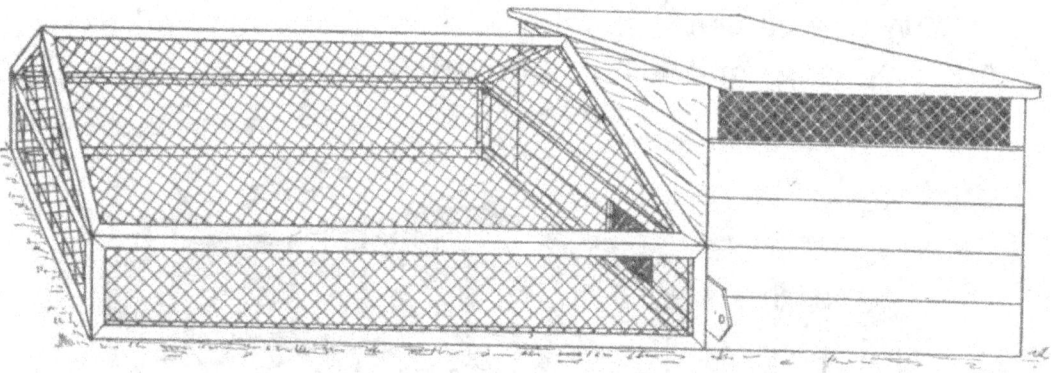

**ANOTHER GOOD BROODER COOP.**

## TO THE WIVES OF THE FARMERS.

I have written a great many articles about the great importance of the poultry interest to the farmers of the South, and while there is now going on all over the South a great awakening among the farmers, it seems that this great interest should show a greater awakening than it does, so I am going to have something to say to their wives. It is said that upon one occasion a farmer's wife was set upon having a certain thing done that her husband opposed. He finally said: "I want to tell you that I am the head of this family and it shall not be done." "Well," said the good wife, "you are surely the head, but I am the neck, and the neck always moves the head as it wishes," and the head moved that way. So I am going to write one or two articles to the necks. Women have a natural God-given talent for breeding poultry, and where they read up on the new methods of breeding poultry, raising early chickens and learn how to prepare eggs so as to market them to the very best advantage, it is wonderful what a success they meet with. I read often in the papers how intelligently they discuss the problems that are under discussion among the poultrymen, and then at the great success that they meet with in the show room. I stood by and heard it asserted that a woman was offered one thousand dollars for five Buff Orpingtons that she had at the Atlanta show, and she had been in the business for only five years. Now, this is (I admit) a somewhat extreme case; but with eggs at forty cents a dozen for case eggs, and for eggs properly packed, clean, as nearly as possible all of one color, bringing in the cities and towns fifty cents, and hens at seventy-five cents each, with no broilers and fries to be had at any price in Georgia, in the early spring, there is surely something wrong among the farmers that their wives should correct, and if they do not know how to turn the trick, they should quickly learn how by reading up on poultry; get in a stock of the best breeds and thereby keep some of the money at home that the Tennessee folks are pulling out of our State.

## FARMERS' WIVES AND COLD STORAGE.

Did anyone stop to ask why eggs are put on cold storage? Well, because there are not enough eggs produced in the United States in the winter to half supply the demand, and so in the spring eggs are bought up in large quantities, put on cold storage and held over to supply the winter demand. This is a fact that should be known by every farmer's wife in the South, for in the North, while they do produce large quantities of winter-laid eggs, they do it at so much expense that it strikes the profits a heavy blow. What an enormous advantage the farmer's wives have here if they would take advantage of this fact and learn some simple things about poultry culture! I believe that it would be a paying investment for the legislature to provide for a poultry bureau under the control of the agricultural department, empowering the department to send a well informed Southern poultry man to lecture to the farmers' wives on poultry culture, and especially on the production of winter-laid eggs, and how to increase the supply. The Southern farmers' wives, properly instructed, could in a few years produce a sufficient quantity of eggs to run cold storage eggs out of the Southern market and put millions of dollars into their pockets.

In our splendid climate we have the ability to grow winter and summer every article of food that is necessary to produce an abundant crop of winter-laid eggs, and this would mean twice as many winter-laid eggs as the entire country could consume. We would then be sellers and not buyers of winter eggs. This will never be accomplished, though, until the farmers' wives take the matter in hand. Think what a happy day that would be! Bill out in the field fixing for cotton, tugging at the bell rope with which he was gee-hawing old Balaam, and Mary in the poultry-yard singing, to the sweet music furnished as an accompaniment by the cackle of the hens, and thinking of the happy time when she will market the eggs and have the dollars rolling into her pocket.

In 1903, the Western and the New England States awoke, as from a dream, to the importance of the egg business in the United States. Previous to 1903 the papers, poultry journals, farm papers, and the dailies commenced writing up its importance as a money-making business. New England was the first to give the business much attention; the West—more particularly the far West—opened their eyes and went after the dollars that they have since gathered. Then there was an awakening that traveled south along the Mississippi valley. The business increased with wonderful rapidity.

Kansas, in 1903, produced $6,498,856 worth of eggs, and in 1907 the output was $10,300,082. Today the bulk of our eggs comes from the farms in the Mississippi valley. Unfortunately, it has not spread farther over the South. There is, however, a marked increase in the interest that the cotton belt states are taking in the production of eggs and poultry. There is a matter, however, that should attract the attention of the wives of the farmers, and that is the quality of the eggs produced on Southern farms, and also the great loss that they sustain in the careless manner in which they handle them. The eggs produced by the common or dunghill hens that are bred all over the South weigh nineteen and one-half ounces (average) to the dozen; the average Tennessee weight is twenty-one and one-half ounces to the dozen; Western eggs, from twenty-four to twenty-five and one-half ounces to the dozen. The latter are from well bred fowls.

Now, the above facts show, even without the application of the up-to-date methods of handling improved poultry, their great superiority over the common stock in the matter of producing heavier eggs. In the Eastern States and largely in the West the common hen is a thing of the past. In the Eastern and Western States and largely in the Mississippi valley the farmers avoid letting their hens lay around the barn or in the fence corners or in the weeds, because they can not ship soiled or stained eggs. In many localities they are unsalable, so they provide nice clean nests, and, when necessary, force their hens to lay in them. A soiled or stained egg is frequently unfit for food, for eggs quickly absorb dampness and with its odors and uncleanliness that affect the taste and quality.

Pack a few eggs in pine sawdust. In two hours break one of them and you smell turpentine.

The following seven breeds of fowls are, in the opinion of the majority of the poultry breeders of the South, best adapted to our climate: Orpingtons, Rhode Island Reds, Plymouth Rocks, Wyandottes, Langshans, Leghorns and Minorcas. The first five named have more or less Asiatic blood in their veins. The two latter are of Mediterranean origin. The five first named generally lay dark colored eggs. The two latter lay clear white eggs. The Minorcas lay perhaps a larger egg than any other breed. The Leghorns and Minorcas require warm, comfortable winter quarters, and if well cared for they will lay fairly well in the winter. The other varieties should not be kept too closely housed, but they must be kept where the cold north winds can not reach them in their roosting house. The following suggestions as to the general treatment of fowls in order to increase the winter production of eggs may help you:

Provide comfortable winter quarters, facing south. Do not let them roost in the barn or out on trees. Do not house the first five breeds in quarters that are too warm. The two last named must be kept comfortable at night and require warmer quarters at night than the others. Feed them regularly, and let them go to bed with full crops. Keep clean, fresh water before them at all times. Gather the eggs once a day, and in warm weather twice a day. Never take an imperfect or soiled egg to market. In the summer place the eggs in a cool, dry place, where there are no offensive odors. Avoid straw or hay for nests; they produce mites. Use excelsior if obtainable; if not, use clean, dry, fine sand or earth that has never been fertilized.

Now, I have said some of these things before in my articles, but not to the wives of the farmers, and I close these letters to them with the wish that every farmer's wife in Georgia would commence now to take a deeper interest in poultry culture. Select the breed that seems best to you, read the poultry papers, study this book, make a wired poultry run, and you will soon be glad that Uncle Dudley advised you to breed poultry.

## STRAIN.

To those who have recently commenced to breed poultry in the South—"and their name is legion"—there are many terms used by poultry men that they do not clearly understand.

I have been frequently asked the question, "What is meant by utility fowl?" The answer is, of course, a general purpose bird of pure blood, but a little off in shape, color, comb, legs or eyes—a bird that is as pure in blood as the blue ribbon bird and that will lay as many eggs, but that will not fill the requirements of the standard of perfection. The above is easily understood, and there are comparatively few now who do not know what is meant by a utility bird. There is another term used among poultry men that many of those who are new in the business do not fully understand, and that term is "strain."

Everybody knows that "strain," when applied to poultry, means a breed that has been brought up by skillful breeding to a greater degree of perfection than those of the same breed, that are raised by other breeders.

Some one has said that "strain is as important as breed," and this is really a fact. It takes as much knowledge and experience, together with good judgment, to improve a breed of fowls as it does to originate them. It is a practical impossibility to bring a breed of fowls up to the point where, by actual proof of its superiority to other fowls of the same breed, it can justly be claimed as your strain in less time than something like eight or ten years. True it is, that you can make a great improvement in them in three or four years, but will they hold that improvement unless the same methods by which this improvement has been accomplished is kept up? I know that it will not, if these fowls are turned over to others, who follow other methods than those employed by the originator of the strain. Now, had the original methods been kept up for, say ten years, the characteristics that made them superior to fowls of the same breed would have been so strongly fixed that their chicks would almost all of them continue to breed after their parents. Now, how is a strain formed? The Standard gives us a description of a positively perfect bird, but it fails to make any mention of the number of eggs that each bird should produce; this, of course, could not be done in the Standard, but outside of the Standard (which only tells you how a bird should be shaped, how the feathers should be colored, and where each color should be found on the bird, how the eyes, legs and toes should look), not only that they might be able to win blue ribbons at the show, but also that their shape should be built up so that they would best do that which they were bred especially to do, i. e. to produce eggs or meat for market.

Now, the man who wishes to establish a "strain" has to have a perfect knowledge of the requirements of the Standard as to the breed that he is trying to improve. Beyond the Standard he has to improve the egg product and at the same time, build up the constitution of the bird.

You see then, that after this bird has been well nigh perfected by skillful breeding, that the same degree of skill is required to enable you to keep it up to a degree of perfection that you found in it when you purchased it. It is no small matter to do perfectly the three things as indicated above. So many things have to be held constantly in the mind at once, so many small (and large) details have to be done at the proper time, that any one who is at all careless, or forgetful, is apt to let some of them go undone, when the result would be only a partial success. A pen of the most perfect birds that were ever produced, will not produce chickens that will all be as perfect as their parents. Your business will be—if you purchase eggs from this pen—to select from the chickens, those that bred together would produce the largest percentage of fowls like their parents. It takes good judgment, backed up by long experience to do this.

Then when you have made a success here—you may have fallen short on the egg product or in their vigor or healthfulness. Hens that lay the largest number of eggs do not generally produce the strongest chickens constitutionally, and yet the healthiest and most vigorous pullets usually make the best layers. Again, a cockerel may be perfect in every way except in vigor, and yet produce weak, imperfect chickens. A less perfect, but more vigorous, cockerel would produce a larger percentage of perfect chickens.

These are only a few of the points that you would have to understand in order to enable you to hold a perfected strain up to the degree of perfection in which you secured it and to which its originator brought it.

## SHAPE IN POULTRY.

For the past two years I have been writing almost entirely to those who were just beginning to breed poultry in the South. The folks who did not know how to incubate eggs—how to feed the little chicks and how to care for them generally until they come to maturity.

There are many things that those who have just caught the chicken fever, and also those who have had this disease long enough for it to have become chronic, must learn beyond the above first lesson.

One of the things that is highly important is the shape that must be carefully adhered to in each breed. Every breed of thoroughbred poultry has a shape that is distinctly its own. It is different from that of every other breed and is of more importance than the feathers, from a commercial standpoint.

Every one of the different American breeds that we have with us today, were built up for a special purpose, In many instances it took long years of constant watchfulness and labor by skilled breeders to produce results that made each breed valuable, because it did just what it was shaped up to do, and because it did just what other breeds, though just as valuable, could not do as perfectly as this breed could do.

Now there are several breeds that are not mixed breeds. They are those that have come to us from foreign lands. These fowls come to us from countries where they have been bred for many long years, just as they are today. They are known and classified generally by the names of the country where they were in the long ago originated and they are known as the unmixed breeds—the Mediterranean from southern Europe, the Asiatic from China, the Malays and Sumatras from the South Pacific Islands, the English from England and the French from France.

It is from these breeds, crossed together, that we have the American breeds. The foreign breeds all have a shape peculiar to themselves and therefore the American breeds take their shape generally from the amount of foreign blood that predominates in their veins.

American breeders have always been rather "luni" on the subject of creating new breeds, and yet when not carried to excess, as has been the case in late years, there has been "method in their madness."

The new breeds that have been originated by American poultrymen are, if not superior, equal to those of any other country, but the almost innumerable varieties of these distinct breeds is where the "luni" phase of their work comes in.

Now, as I have said, each of the American breeds have a shape that is different from all other breeds, and the variety of each breed, although entirely different in color, must conform strictly to the shape of the original breed for which it is named.

There are eight varieties of Wyandottes, all different in feather, but all alike in shape. The Asiatics are shorter in their backs and bodies, and the Mediterraneans are longer than other breeds. The American breeds are in this respect, built up between these two extremes.

It is therefore, absolutely necessary in breeding poultry, to study closely the shape of the breed that you are handling, and so mate them together that they will produce chicks that will conform to the requirements of the standard of perfection.

CORRECT WYANDOTTE SHAPE.

## STANDARD WEIGHTS OF FOWLS.

We have often been requested to publish the standard weights of the various breeds. It is a good idea to have the correct weights of your breed in mind when purchasing.

| Breed. | Ck. | Ckl. | Hen. | Pul. |
|---|---|---|---|---|
| Plymouth Rocks, all varieties | 9½ | 8 | 7½ | 6½ |
| Wyandottes, all varieties | 8½ | 7½ | 6½ | 5½ |
| Rhode Island Reds, all varieties | 8½ | 7½ | 6½ | 5 |
| Light Brahmas | 12 | 10 | 9½ | 8 |
| Dark Brahmas | 11 | 9 | 8½ | 7 |
| Cochins, all varieties | 11 | 9 | 8½ | 7 |
| Orpingtons, all varieties | 10 | 8½ | 8 | 7 |
| S. C. Black Minorcas | 9 | 7½ | 7½ | 6½ |
| R. C. Black Minorcas | 8 | 6½ | 6½ | 5½ |
| S. C. White Minorcas (no rose combs) | 8 | 6½ | 6½ | 5½ |
| Langshans | 10 | 8 | 7 | 6 |
| Dominiques | 8 | 7 | 6 | 5 |
| Javas | 9½ | 8 | 7½ | 6½ |
| Buckeyes | 9 | 8 | 6 | 5 |
| Silver Gray Dorkings | 8 | 7 | 6½ | 5½ |
| White Dorkings | 7½ | 6½ | 6 | 5 |
| Cornish Fowls (Indian Games) | 9 | 7½ | 6½ | 5½ |
| White Faced Black Spanish | 8 | 6½ | 6½ | 5½ |
| Houdans | 7 | 6 | 6 | 5 |
| Bronze Turkeys | 36 | 25 | 20 | 16 |
| Pekin Ducks | 8 | 7 | 7 | 6 |

The Leghorns, Anconas, Hamburgs and Polish have no standard weights.

A cock is a male bird over one year of age.

A cockerel is a male bird under one year.

A hen is a female bird over one year of age.

A pullet is a female bird under one year of age.

## A HISTORY OF SOME OF THE BREEDS OF FOWLS RECOGNIZED BY THE STANDARD OF PERFECTION THAT ARE USED FOR COMMERCIAL PURPOSES.

The "Standard of Perfection" is issued by the American Poultry Association for the use of poultrymen in the United States about once in four years. It gives a complete description of each breed, as to name, shape and color. It describes the ideal fowl, the absolutely perfect bird, a bird that is only described as a standard by which the poultryman is to work, in order that he may breed his fowls up to the point of the perfect bird as it is described. This perfection has in no case ever been produced in a fowl. Some specimens have been produced that come only a few points below the score of 100, which means perfection, and to which no bird has as yet attained. Now we find in the Standard only seven different classes of birds that are used strictly as utility birds:—the American, the English, the French, the Mediterranean, the Asiatic, the Oriental and the Hamburgs.

The American class is represented by six breeds: Plymouth Rocks, Wyandottes, Javas, Dominique, Rhode Island Reds and Buckeyes.

The English by three breeds: Dorkins, Redcaps and Orpingtons.

The French by three breeds: Houdans, Crevecoeurs, La Fleche.

The Mediterranean class by five breeds: Leghorns, Minorcas, Spanish, Blue Andelusians, Anconas.

The Asiatic class by three breeds: Brahmas, Cochins, Langshans.

The Oriental class by three breeds: Cornish, Sumatras, Malays.

The Hamburgs are in a class by themselves—the Dutch class.

I propose to give a short history and description of the majority of the above breeds. I will describe where it is possible to do so, their origin in this and in other countries. In the American and English breeds, how the blood of different breeds were mingled together in order to produce certain characteristics that differentiate them from other breeds.

I think that because of the great interest that is now being manifested by people all over the entire South in poultry, that these articles will be not only interesting but instructive to those who are in any way concerned in one of the most interesting and profitable industries of our God-blessed country. Few of us as yet have realized fully the enormous possibilities of poultry culture in the United States and particularly in our Sunny Southland. When we stop for a moment and think that but a few years ago this neglected industry amounted to so little from a commercial standpoint that it was scarcely recognized by the agricultural departments of the different states, or of the department at Washington as an agricultural product and that today it has suddenly, as if by a miracle, grown to such enormous proportions. We are simply amazed that the people everywhere were so slow in realizing its importance and its vast possibilities. It was only about twenty years ago that the poultry product of the United States amounted to about $250,000,000. In ten years it went to about $700,000,000 and we are informed that in the next census about to be issued by the government, that the value of this hitherto unimportant product will exceed in value $1,000,000,000. I will give a description of the American breeds of fowls that together with the foreign breeds have made it possible to bring about this transformation of the poultry interests in the United States.

## THE ASIATICS.

### THE SHANGHAI FOWL.

It was in the middle of the fifties that I saw for the first time in my life a dark, terra-cotta colored egg. An uncle who was fond of poultry, gave twenty-five dollars for a trio of a new breed of fowls that had been brought from China some years previously. I have a distinct recollection of these fowls. They were light colored, or white with dark markings. They were very large, with long, feathered legs, and the person from whom they were purchased said that when fully grown the male bird would be able to eat off of a flour barrel. Many other wonderful things were told of these remarkable fowls. Previous to the advent of the Shanghai, the common chicken in South Carolina was largely mixed with game, and because of their quick feathering, particularly their wing feathers, it was difficult to raise them. Now, of course, after the excitement had cooled down and many others commenced to breed the Shanghais, I was given a setting of the eggs. I think my chicken fever, which has since become chronic, began with that setting of eggs. Well, the chicks hatched and commenced to grow, and, wonderful to behold, instead of being heavily feathered, they were over half grown before they commenced to feather at all. There was a feeble effort made to feather, but they were nearly naked. Now, they were quick growing, and very hardy, and when fully feathered very beautiful.

The above is my boyish recollection of the Shanghai fowl. After this I remember the Brahma Pootra, and the Chittagongs, and then came the Cochin. I have written the foregoing as the starting point in the history of the

Asiatic fowl, for my purpose is to write the history—or what is known of the history of the most wonderful, to my mind, fowl on earth. They have been traced back in China—some of the strains—for nearly two thousand years. I have read accounts of the origin of these fowls that are somewhat confusing, for these accounts tell us that the Shanghai and the Cochin are identical. This is an error. The Brahma that we have today came from the old Shanghai, while the Cochin is of a different strain. The Buff and Partridge Cochins have been bred in China for hundreds of years—somewhat like those bred in America today.

No breed of fowls in America—or, for that matter, in England—has as interesting history as has the Asiatic fowl. It was early in the fifties that a vessel arrived in New York from Shanghai with a large number of fowls of different colors. They were larger than any yet seen in the United States. Not only this, but in many other respects, these fowls were entirely different from any fowl that Americans had ever seen. George P. Burnham bought from the captain of a vessel a number of them that were gray in color, and commenced breeding them. There is no doubt that there were a few brought over before. A missionary brought some of them from China to his father, who lived in Connecticut, in 1847, and then a few were received in 1849. But Mr. Burnham sent nine of his flock to England, presenting them to Queen Victoria in 1852. This attracted the attention of the whole country to them. The only breed that had been given any attention before this was the game. They were bred all over the country by aristocrat and plebeian. They were bred for no other purpose than to fight. News traveled slowly in those days, but in comparatively short time the wildest rumors spread over the country concerning the Shanghai fowl. Fabulous tales were told about them.

Everybody talked of them, and as the reports traveled over the country they rivaled (if they did not surpass) the history of the three black crows. Enormous prices were paid for them, but not as large as the yarns that were spun concerning them. The record as to many things pertaining to their history in the early fifties has long been in dispute, but the record is clear as to one thing, and that is, that the first case of chicken fever occurred in the early fifties; that it spread rapidly, became epidemic, and that m men, women and children have it to-day than ever, and the number is still increasing. To the old gray Shanghai belongs the honor of being father of fancy chicken breeding in these United States.

## DE SHANGHI CHICKEN.

*Oh! de Shanghi Chicken*
  *Am a mighty funny fowl,*
*Said de double-headed Pigeon*
  *To de one-eyed Owl;*

*De old Grey Goose,*
  *Wid de web 'tween her toes,*
*Most kills herself a laffin*
  *When de Shanghi crows.*

*De Shanghi Rooster, he gro so hi*
  *He head it almost tech de ski,*
*And when dat Rooster gins to cro*
  *He neck bens back des like a bo.*

  CHORUS:
*Oh Shanghi! Shanghi*
  *Don't bet your money on de Shanghi.*
*Catch a little chicken; drap him in de ring,*
  *But don't bet your money on de Shanghi.*

To-day two breeds represent the Brahma-Pootra, Cochin-China, Chittagongs, gray Shanghais and some other names by which they were known in the fifties, and they are all known now as the Brahmas and the Cochins. I will take each of them up in another article.

Years ago there lived on a plantation near Summerville, S. C., two old maiden ladies. With the exception of a nephew, whom they had adopted and reared, their relatives were all dead. The boy, whose name was William Smith, was a bright, energetic, good fellow, bubbling over at all times with good will for all mankind, and fun for his companions. Now, the two old ladies were old-time Methodists, and like all of the old Methodist folks, kept open house for the preachers. Among the preachers was one, a grand, good old patriarch, who used to say that he was at home in South Carolina, wherever he had his hat on. His name was "Uncle" Sidi Brown. Billy Smith used to say: "Yes, and more so, when it is hanging up on our hatrack." Well, Billy said that upon one occasion Uncle Sidi stopped at the gate at about eleven o'clock in the day, got out and came in. He received a cordial welcome, his horse was "put up" and the good old folks set about getting dinner ready. The chicken cholera had just swept over the country and all that was left these good old folks was a Shanghai rooster, a guinea hen and a Muscovey drake, and Billy was told to catch one of these for dinner. After a vain effort to locate either of them, he returned and reported the fact to his aunts. The old ladies were forced to sit Uncle Sidi down to a dinner consisting of corn-bread, bacon and greens. Uncle Sidi enjoyed his dinner and then called for his horse. He bade the folks good-bye and started out. Just here the old Shanghai appeared in the backyard and crowed: "Sidi is g-o-n-e!" but Uncle Sidi had forgotten his umbrella and started back for it. Now the guinea hen was doing picket duty on the fence, and as Uncle Sidi started back, gave the alarm: "He's come back! He's come back," and then old Muscovey sounded a warning, when he stuck his head from under the house and said: "Hush-hush-hush," and so ended what came near resulting in the tragic death of one of the three that survived the cholera, and thus endeth the history of the Shanghai fowls.

## THE BRAHMA FOWL.

The Brahma and the Cochin fowls that are bred in large numbers in the North to-day and that are so highly esteemed there, as I have said in a former article, are the descendants of the fowls that in the latter part of the forties and in the first of the fifties arrived in this country and created a great sensation. We are told by travelers, missionaries, and others that in the interior of China the people are experts in incubation and breeding poultry, and that the fowls differ in size, shape and color, according to the location in which they are bred, and that they are taken to the nearest city for sale. The original importations were generally named for the city that they came from, hence Mr. George P. Burnham selected and bought light-colored fowls from the captain of a vessel that came from Shanghai and named them Gray Shanghais, and there is no doubt in my mind that these fowls are the progenitors of the Light Brahma of to-day, but this is disputed. Of course there may have been some admixture of other Asiatic blood, but in shape and color the fowls that I bred as a boy resembled the Light Brahma of to-day very closely, as I remember them. The dark Brahma resembles the light or penciled Brahma in many respects, and evidently came from dark-colored Shanghai stock. The Light Brahma is the largest fowl of which we have any knowledge, and weighs one pound more than the dark

Brahma, both cock and hen. Their weight by the standard is, cock twelve pounds, hen nine and one-half, while the dark Brahma cock should weigh eleven pounds, hen eight and one-half. Very few of the Brahmas are now bred in the South. They were once very popular. Those who once bred them found them sluggish. They are not good foragers. Their great size and consequently their weight, made them poor mothers and they were cast aside for the thrifty Leghorns and the barnyard hens were used for mothers until the Plymouth Rocks came, and largely supplanted the Leghorns. In the New England States they are still highly esteemed, particularly the Light Brahma. Both the light and dark Brahmas are good winter layers. The chickens, it is claimed, are more easily raised than those of any other breed. The light variety lays next to the largest egg of any other breed, and, it is claimed, heavier than all other eggs. The Minorca lays the largest egg of any breed. The dark Brahma would be more popular among breeders but for the difficulty in getting them bred to standard requirements. The Brahma, however, still stands at the head of all fancy poultry bred in America, and will perhaps reign king for many years to come.

## THE COCHIN FOWL.

In the article that precedes this, I have said much about the Brahmas that can also be said of the Cochins. In fact their history in America is so interwoven and, as it were, blended together that breeders have been squabbling for about forty or fifty years in the honest effort to get their history correctly written without having been able to straighten it out. A number of books have been written pro and con on the subject. Travelers have been interviewed; men going to different parts of China have been asked to investigate and report on their return. Then various reports have been discussed in the poultry papers, books written about them, and then discussed in the clubs, and to-day we are just where we commenced. It would fill several pages of this book to notice a small part of them. The claim on the one hand is that the Cochins and the Shanghai are identical and on the other that they are a separate and distinct breed. The Cochin family, it is claimed by some, came from Cochin, China, and while this is disputed, the evidence is largely in favor of the first and against the latter claim. It is uncertain when they first came to America. It is supposed that they came over in the early fifties and were called Shanghais. They evidently are not Shanghais, for they are nothing like the latter in shape or feather. A missionary who traveled years ago in China and India mentioned the fact in speaking of the dense ignorance of the people—that the people believed that the spirits of their dead went into fowls that were a beautiful buff and others of them, like the Partridge Cochin of to-day. In his description of the fowls he stated that their size was so great because of the care that the people took of them. The record of them, which was carefully kept, went back about fifteen hundred years. Now there are four varieties of Cochins, Buff, Partridge, White and Black, alike in every respect except in color They are shorter in body than the Brahma, legs shorter and more heavily feathered. The standard weight is the same as that of the Dark Brahma. It is claimed by many that they are better winter layers than the Brahmas; that their chickens mature more quickly, and that they are better mothers. However, these characteristics largely depend on the strain from which they come, and the man behind the guns.

## THE LANGSHAN.

Here we have a fowl with a history which, unlike the other Asiatics, is pretty well known, and over which there has been very little dispute. There are two varieties of this breed—black and white. They were brought from China by an English officer, Major Croad, in 1872. There was a difference of opinion at the time as to whether or not they were a new breed; but this, upon a close examination, was soon adjusted, for it was made clear that they were a distinct breed and that they had been bred in China, in and around the Langshan district, for generations back, and also that they were almost regarded as a sacred bird by the Chinese who bred them, and who carefully kept them from admixture with other breeds.

It was at first claimed that they were Cochins, and to a novice the black Langshan and the black Cochin look alike, but there are many differences between these fowls. In shape, they are totally different.

They have given, and are now giving better results as a utility fowl than any of the Asiatics, and in some respects, than the American-bred fowl. They stand our climate better than the other Asiatics and are therefore less liable to diseases. They are not lazy, as are all of the other Asiatics, but are energetic and will get everything that any fowl could on a range, yet they are quiet and gentle, and do well in confinement. Their standard weight is nine and one-half pounds, cock; seven and one-half, hen. They are excellent mothers, and great layers, and are as good or better producers of winter eggs than any other breed. Their chickens mature, it is claimed, by those who breed them, more quickly than any other breed. They are vigorous and strong and not as liable to disease as some of the other breeds.

They can stand any amount of cold as they are heavily

feathered, and yet are not affected by the heat of our summers, as are the other Asiatics.

They do better for the amateur than any other breed, and have the ability of taking care of themselves, and of shelling out their rich, dark large eggs on almost any kind of food if you will only give them enough of it with grit and fresh, clean water.

They are of fine carriage, a grand, aristocratic-looking fowl in appearance, more symmetrical in their make-up than their Asiatic kinsman.

Their flesh is white, and of fine quality. Their legs are not heavily feathered and should be of bluish-black, without a shade of yellow. In color the blacks should be a bright, slightly bluish black, with no semblance of purple sheen. Their legs are longer than any of the Asiatics, as also are their backs and tails. They are rather full in the breast. These fowls bid fair to become in the near future one of the most popular breeds as a utility fowl in the South.

# THE AMERICAN BREEDS.

## THE PLYMOUTH ROCK.

No one can dispute the fact that since the Plymouth Rock made its bow to the public it has steadily advanced in favor with the poultrymen and the farmer. To-day it is the most popular fowl bred in America. The Shanghai, Brahma, and Cochin fever has cooled. Their slothfulness and the activity of the Leghorn had much to do with the bringing about of this result, for when at this juncture this fowl made its appearance it supplied a want that existed all over the country. The Asiatics were too slow, but the Mediterraneans did not sit; so when a fowl appeared that had the best characteristics of both of the above breeds, it at once filled a long-felt want, and at once took a strong hold on all classes who were interested in poultry. Its reception when, it became known, was almost as enthusiastic as that of the old Shanghai. Folks went wild over it, and told wonderful tales as to its superiority over every other breed. It could never have maintained its hold upon the people had it not proved its sterling good qualities. It had defects, which I will not enter into, for some of them are slight, and its good qualities cover the others to such a large degree that they can be overlooked. They are gentle, of robust constitution, easily acclimated. They seem perfectly at home from Canada to Texas. They seem to thrive in the backyard of a town lot or on a good range on the farm. Their chickens are generally strong and easy to raise. They mature quickly; are good winter layers (although in the two latter points it is claimed that some of the new breeds surpass them). They are superior utility fowls. There is actually no certainty as to their "make-up." It is claimed that they originated in Worcester, Conn., and that they were the result of a cross of the old American Dominique and Black Cochin. This is disputed by prominent breeders, who claim that they are the result of a cross of black Spanish and white Cochin, and the fowls of this cross bred on the American Dominique again; that they are the result of a blood mixture of the Dominique, black Java, light Brahma, dark Brahma and pit game. They made their first appearance in 1869, and attracted very little attention. Afterward, there were some six or more claims as to the breeds that composed their foundation stock; but out of it all came the final result—the production of a grand, good utility fowl, the first breed that was ever built up by American breeders. There are six breeds recognized by the standard—barred, white, buff, Columbian, partridge and silver penciled.

The fowl bore the name of Plymouth Rock until the advent of the other varieties, when "Barred" was added in order to distinguish it from the other varieties. We

A WHITE PLYMOUTH ROCK COCKEREL
OF CORRECT CARRIAGE.

are told that about the year 1876, Oscar W. Frost, of Monmouth, Maine, had hatched some white chickens from Barred Plymouth Rock eggs. He raised them and found that they had the Rock shape and characteristics. Then others reported the same results from Rock eggs. In tracing the breeding back of these white sports, it was found that the strain from which they came had been crossed on a fowl called White Birmingham. Several other whites came out in this strain, but finally these different strains were bred together and resulted in the White Plymouth Rock of to-day—an exact counterpart of the Barred in everything except color.

In 1890 the first Buffs were exhibited. It seems somewhat singular that in this year, or the year after, Rhode Island Reds and Buff Wyandottes made their first appearance.

Originally there were two separate strains produced at the same time—the Wilson strain and the Buffington strain. The Wilson strain was the result of a cross between Buff Cochins and Light Brahmas. The Buffington strain was made up from a cross of Rhode Island Reds (as yet an unknown fowl) and White Rocks. Neither strain was a perfect Rock type, but both had some strong, well-defined Rock characteristics. These two strains were bred together and we have to-day the Buff Plymouth Rock that is very rapidly forging its way to the top in this trio of popular, all-purpose fowls.

## THE RHODE ISLAND RED.

No American breed ever came as quickly into popularity as did the Rhode Island Red. The Plymouth Rock made its appearance several years before it took hold on the public. In fact it was a failure when first introduced. It was turned down—sent back, had to be built over and then on its second appearance became popular. The Wyandotte was three years trying to get into the standard, but this fowl, bred for years in Rhode Island, was an unknown commodity beyond Rhode Island and the adjoining States. It attracted the attention of a few poultrymen and from the start jumped into popularity even before it was bred to anything like perfection. No one knows how or when they originated. As bred in Rhode Island, it was only true to shape. Many of them were red, hence the name, but they were buff and brown, and other colors among them. There was for years no effort made to breed them to color. They were only bred as utility fowls. Dr. N. B. Aldrich, of Massachusetts, has traced them back for sixty years. He is of the opinion that they came from the Red Cochin, which is probably correct, but what other breeds that were crossed on the Cochin to produce this fowl as we find it to-day, is very uncertain. For fifty years it has been in process of formation, and there is to-day no breed of fowls that has been produced as has this breed, by out-breeding entirely. They were first exhibited in New York City in 1891-92. They were admitted into the standard in 1904 since which time they have been largely inbred. Previous to 1901 they were bred by many poultrymen according to their own judgment and by the introduction of blood from almost every breed that had red feathers and others that were of other colors, but they were almost unanimous in their desire and purpose to use nothing but Red cock birds. They are far from the perfection attained by other breeds to-day. They are very difficult to breed true to the requirements of the standard.

There is a tendency to show feathers on their legs, and to show smoot on the under feathers. Then the exact shade of cherry red is hard to produce to what, according to rules that govern mating in other breeds, would bring about the proper result. The standard demands a deeper red in the cock than in the hen. Both however, should have in every feather where red is required red to the meat. Beak and legs should be bone colored. In the single-comb variety, which is more popular than the rose comb, the comb should have fine, well-developed points; the eyes red, face bright red, wattles and ear lobes red. In shape they should be rather long in the back, carried nearly horizontal—breast broad, full and well rounded; body, broad and long, giving the bird an oblong appearance. There is a great tendency in these fowls, particularly in the cocks, to breed short in the back. In breeding them the short-back cocks should be used as utility birds, and not used in pens bred for show purposes.

## WYANDOTTES.

These birds have an intensely interesting history. Books, articles in the poultry papers, innumerable speeches in the poultry associations, an all-night discussion when they were first put forward for admission to the Standard and then the wonderful way in which they were originally bred—all together make their history read like a romance. Now, I can not in this article, of course, give their history in full, for if I did, it would take up several pages of this book, for it is almost as voluminous as the national ode of the Tartars, which is said to have been written on sixteen miles of parchment; so I will just ontline the story.

In 1866, or about that time, a large number of men in several of the Northern States set out to create, by crossing different breeds together, a new breed of fowl. Some of them were working at random; others had a very definite idea of the shape, color and chief characteristics of the bird that they were trying to produce. It is rather remarkable that one in New York, one in Michigan and one in Massachusetts had in their minds, without the knowledge of the others, very nearly what the others were driving at. The others seemed to be working at random. They each produced a new breed and each had a name for the fowl that he had bred. They were named Hambrights, Hameltonians, Eurekas, Excelsiors, Columbians, Seabright Brahmas, American Seabrights, and perhaps some other names. These different breeds resembled each other in some respects. Now, the fowl that all of these men were trying to produce was a smaller fowl than the Asiatic, and a larger fowl than the smaller breeds; so that it would fill a much-needed want in the retail trade, and then at the same time a fowl that would mature quickly, lay large marketable eggs and plenty of them. There was a long, hard fight to get them in the standard, but there was such a great difference in the fowls presented that neither the breeders nor the American Poultry

PRIZE WINNING WYANDOTTE PULLETT AT
AN ATLANTA SHOW.
(From an untouched photograph.)

Association could agree, and they were turned down; nor could they agree on a name. In 1883 another effort was made among the breeders, and they came near another disagreement as to a name, when F. A. Hondlette, of Massachusetts, suggested "Wyandotte." This being accepted and other differences adjusted, they were admitted into the standard. They at once became very popular. Now, these fowls were the silver Wyandottes of to-day. Since their admission, seven other varieties have been admitted, white, buff, black, Columbian, silver penciled, partridge and golden. It is a little singular that these fowls were bred by crossing almost every then-known breed in America—three varieties of Cochins, two varieties of Hamburgs, two varieties of Brahmas, French Breda, the old Chittagong and the little Seabright bantam are among their ancestors, and perhaps some others.

You see advertised in the papers and placarded on the coops in the show rooms sometimes, "Silver Laced Wyandottes." There is no such variety named in the Standard.

Now, all of the above varieties have the Wyandotte shape and weights. Cock, $8\frac{1}{2}$ pounds; hens, $6\frac{1}{2}$ pounds; cockerels, $7\frac{1}{2}$ pounds; pullet, $5\frac{1}{2}$ pounds. They differ only in color.

One of the most important things in all breeds of poultry is their shape; a Wyandotte has a shape peculiar to itself; no other breed is shaped like it; it has a very short back, very broad across the back, a full, well-rounded breast, very small bones. It is said of them that if you separate the meat from the bones and weigh each separately you will find that they have more meat and less bones —in weight—than any other breed and from my experience I am quite sure that this is true. They are excellent mothers, very gentle and quiet in disposition, are easily broken from the nest when broody. They are superior winter layers. Their chicks are healthy and vigorous and mature, it is claimed, more rapidly than any other breed except the Langshans.

They are, besides, very energetic, and do well if confined to small quarters.

This is a very valuable fowl and should be bred more extensively in the South than it is. Recently the indications are that it is rapidly coming back and will soon regain its former popularity with the poultrymen.

## THE GREAT AMERICAN DOMINIQUE AND THE BUCKEYE.

I have previously given a short history of four of the six pure American breeds of fancy fowls. The other two breeds are the American Dominique and the Buckeye. The latter will require but a few words to tell all that is necessary to give their history.

The Buckeyes were originated by Mrs. Metcalf, of Ohio. They were skilfully bred. Mrs. Metcalf had in her mind her ideal of what a utility fowl should be. She made crosses with four breeds—Black-Breasted Red Games, Plymouth Rocks, Buff Cochins and Cornish Game. She bred the result of these crosses together and produced a fowl that in many respects resembled the Rhode Island Red. They were a deep cherry red, with black showing in wings and tail, yellow skin, legs and beak, though some times legs, beaks and toes were horn color. In shape they resembled the Cornish Game, with shorter legs, and feathers were fluffy. Their weight is nine pounds; Cock, eight pounds; Cockerel six pounds; pullet five pounds. They do not measure up to some of the new breeds in eggs or early maturity. They are not largely used by poultrymen.

As to the American Dominique they are undoubtedly the oldest of all American breeds, and were extensively and profitably bred in the North before any of the other American breeds came into existence. No one knows how or when they were originated, like the Rhode Islands Reds, their origin stands out as an unsolved enigma. They are free from any Asiatic characteristics. They were here when old Shanghai made his appearance. They have never been caught with a feathered leg, nor any tendency to have a straight comb. As far back as they can be traced they were scattered all over the country, North and South, before the West was discovered. It is somewhat singular that this breed were very nearly alike in all sections of the United States. Like the unknown facts in the history of most of the American breeds, several men have appeared in print who were "absolutely sure" that they have figured out their origin, but the others who "felt sure" that their theories were correct, differed so radically from the others that at a glance you could plainly discover that all of their theories were based almost entirely on guess-work. Of all the opinions that I have read concerning the history of these fowls, the following seems to me to be the most reasonable, and is probably very near the truth in regard to the origin of this remarkable breed of fowls that now have almost passed away from us. The breed of fowls now known as American Dominiques differed in many respects to those bred sixty or seventy years ago. They were one pound heavier. The color was not exactly the same, nor did they breed out in color as do the Dominique of to-day. In the old stock the cocks were light, and the hens were very dark. In both cock and hen the "Dominique" was clearly defined. In the Dominique of eight or ten years ago this was not the case, and then the weight had fallen off one pound. The old fowl was exceedingly pugnacious. They were almost as game as the game cock. They were strong in wing power and were hard to keep confined. They were in shape somewhat like the Dorking, Rose Comb, ending in a sharp peek behind, only four toes, short legs and long flowing tails. Now, looking at them from the above brief description, Mr. T. F. McGrew of New York, writes interestingly of their supposed origin, when he says, that the Dutch are known to have been the first settlers of New York and that it is certain that they brought with them some of their "everlasting" layers, the pencilled Hamburgs. It is a well-established fact that these Hamburgs mingled with our common white and black fowls will produce a Dominique fowl that the illustrations in the early poultry books show much similarity to the American Dominique, and then the comb, and the feathering, long tail, strong wings and some other characteristics convince him that just there were these fowls first bred and originated. It is much to be regretted that this magnificent breed was allowed to pass from us. True it is, that American Dominiques are bred in this country to-day, but not the old Dominique, that the old people remember. I have a very indistinct recollection of them. I remember that there was such a breed, and that they were bred by some of the planters, who did not allow their negroes to have any other fowls on the plantations. I remember that they were as hardy as an oak, that they had the energy to provide for and take care of themselves. They were great foragers, and roamed over the entire place with almost as much liberty as was given the turkeys. I have read of them, that they were extensively bred in Virginia some years before the Confederate War.

Again, it has come down to us as a tradition, that old Dominique was by our mothers thought to be the top-notch chicken in the barnyard. The great desire for something new among our people has been the prime cause of the passing away of several breeds of fowls, that were not, and to-day have not been replaced by the new breeds. Unquietly think over the past, the present and the future of our poultry interests. I see from my viewpoint that in the great tendency to multiply the breeds of our poultry, we are confronted with one of the greatest evils that we have to deal with.

## THE JAVA.

There are only two varieties of this strictly American fowl recognized in the Standard. The Black and the Mottled. Their history—that is of the pure American Java—is very interesting. They have recently been traced back to 1852. There are very few of the pure-bred Javas in existence to-day. Several breeds have sprung up from time to time that were called Javas. Some of them were admitted into the Standard, and then in a year or two dropped out and disappeared.

It has been claimed that the Plymouth Rock had Java blood in their make-up but on close examination latterly, this proved to be incorrect, for the fowl that was used in this cross proved to have been the Black Cochin; it had feathered legs, and the pure Java has no Asiatic blood in its veins.

Their origin has been carefully traced back to 1852, beyond that no one can tell how they were built up or where they came from. It is said and in fact proved beyond a reasonable doubt, that there lived in Missouri a very prominent doctor who had a breed of black fowls that were all alike in shape, feather and other characteristics that were unlike any fowls then known. They were in shape unlike the Malays or the Asiatics.

They were of large size, long backs, very full breasts, with legs of medium length, and not feathered. They presented a very stylish, beautiful appearance, were full of energy, great layers of large white eggs. They were hardy as chicks, and also in maturity. The doctor who owned them would not sell or give any of them or their eggs to his neighbors. But one of these neighbors, hired the doctor's coachman to give him only three eggs from which he succeeded in raising three chickens. He guarded them with great care and in 1857 moved to Duchess County, New York, and brought a large bunch of the fowls with him. In 1867 they were introduced into Orleans county, New York, where they were bred in large numbers. They have been carefully bred there since that time.

They occasionally cast a white spot, and by this means some think that the mottled variety was produced. It is claimed, however, that the mottled variety was produced by a cross between a Black Java cock and a large white hen that was selected from a large flock of white fowls of unknown breed, and that this was really the way that the mottles were produced, and that because of the fact that the Black Javas had been inbred for many years, this outcross resulted in the product of a fowl, the superior of the blacks in many respects.

The Java—both varieties—is a very superior breed of fowls; and had they been originated in any other country save in the United States, they would be bred in large numbers to-day. The fondness of the production of new breeds among American breeders is, in my opinion, a real injury to the poultry interests. Thousands of new breeds have been produced while hardly one in a thousand have been able to exist. They start up, create a small sentiment and then, when the next makes its appearance, pass out and are forgotten and then again, when one of the new issues gains a footing, and succeeds, it does it, at the expense of an old breed that is in some instances its superior in many respects.

Another evil that should be checked is the multiplication of varieties of the established breeds. Some of these varieties are so nearly alike that good judges can scarcely tell the difference in the show room.

I stood by in a show room last year and saw a prominent judge put a blue ribbon on a coop of Silver Wyandottes that were entered as "Laced" Wyandottes and then on a coop of Pencilled Wyandottes as Silver Wyandottes. He was surprised when I informed him that there was no such breed as a "Laced" Wyandotte in the Standard. It took the book to convince him of his error.

This small difference should be stricken out. They do not amount to anything that is useful and only creates all sorts of confusion. Better drop out the multiplication of the new breeds and the many varieties of the old ones and pay this wasted attention to the utility part of the business. We would thereby have fewer fads and more eggs.

## MEDITERRANEANS.

There are several varieties of fowls known as Mediterraneans—Leghorns, Minorcas, Black Spanish, Andalusians and Anconas. They are all non-sitters. The Leghorn stands at the head of this family as egg-producers. The Minorca is the largest and heaviest. It is generally conceded that the Andalusian is the most beautiful. The Anconas, the smallest of the family, are close competitors of the Leghorns as layers and of the Andalusians in beauty. The latter are not bred to any large extent in this county, while in some parts of Europe they are highly esteemed, being prized because of their beauty and their good qualities as layers. They all lay pure, white eggs, and so did all other fowls in Europe and America until the introduction of old King Shanghai. There is much uncertainty as to the origin of this breed. They have been known for years all over Europe. The Black Spanish have been bred for a long number of years along the Mediterranean border of Spain, and also in the interior, and for many years have been bred just as we find them to-day. There is a great deal of uncertainty as to where the Minorcas originated; many contend that they are identical with the Black Spanish and that they differ only in the white face of the Black Spanish. Others contend that they are a distinct breed that originated on the Island of Minorca in the Mediterranean Sea. There are so many marked differences in the two breeds that the latter opinion is held by a large majority of poultrymen. Very little is known of the history of the Andalusians and also of the Anconas. They have been long known in certain localities of Europe. They, like this entire family, are prolific layers, but differ with the others in plummage. The Andalusians are a beautiful blending of light and dark blue and are generally considered as the most beautiful members of the family. But when you look upon the plumage of the smallest of the family, the Anconas, you are apt to doubt the judgment of others. It is true that, like some of the other members of the family, these fowls in Europe, or in that part of Europe where they are supposed to have originated, are not bred to feather as carefully as are American fowls. The Ancona, best known in America, is thus described in The Standard: Beetle green ground, each feather tipped with white, evenly mottled throughout.

The Leghorns are undobutedly a very popular fowl in America. Some think that they are in advance of the

WHITE LEGHORN COCK.

Plymouth Rock, for lately there has been a larger profit in eggs than in chickens. The Brown Leghorn is undoubtedly the first of the Leghorn family to reach this country. They came in a ship from Leghorn, hence their name. I bred Brown Leghorns for several years before I ever heard of a White Leghorn. I remember that in some journal I saw where some White Leghorns were purchased from the captain of a ship that came from Mediterranean ports. My recollection is that his name was Cook, and soon they were being bred all over the country. This was thirty or thirty-five years ago. It is a well-known fact that the other varieties—black, buff and silver duck wing Leghorns—are the creation of American breeders. The White Leghorns are claimed to be the best layers. The browns are next. The Minorcas lay the largest eggs of any fowl of any breed. The Andalusians lay next to the largest eggs. The Minorcas are the largest and best table fowl and are claimed by some breeders to be the best winter layers. The entire family are not good winter layers, but they can, by careful management, be made to produce a fair quantity of eggs in winter.

## THE BROWN LEGHORN.

Dear Uncle Dudley:

I have been thinking of writing about poultry for quite a while, which I am very much interested in, have been for eight or ten years. I am what you might call a chicken crank sure enough. I read everything in the way of poultry I can get my hands on. I have been in the poultry business about six or seven years only on a small scale, but am gradually increasing my flock. I have been raising the S. C. Brown Leghorns, and think they are the chickens to raise if you want eggs. They are the egg machines. The Barred Plymouth Rocks come next. Some people say the Leghorns don't grow as fast as other breeds and never get large enough to eat. Leghorns look smaller than they really are. Now, if you take two chickens, frying size, one Leghorn and another breed, you will find that the Leghorns have more flesh on their bones than the larger kind. I told a lady some time ago, just to try a broiler and see what she thought of them, so she did, and found them to be just as I described. The Leghorns can not be beaten. They excel all breeds; seem to have more style about them and carry themselves better than any breed I know of. If any one wishes to go into the poultry business I advise you to get the Leghorns, if you want eggs. They are the breed for you; I would not be without them. As to feed, I feed my chickens on most anything. My Leghorns won first prize at the Putnam County Fair in 1910.

MRS. J. A. PINKERTON.

Eatonton, Ga.

P. S. Am anxious to join some poultry association. What do you think about it? P.

---

COMMENT BY THE EDITOR:

This is an interesting letter to me. I bred Brown Leghorns for years. They were then the best egg machines on earth, but a few influential members of the American Poultry Association, for their own profit, crossed these splendid fowls on the Black-Breasted Games, changed their color from brown breasts to black breasts; changed their shape and forced them so changed into the Standard, thereby putting everybody who bred Brown Leghorns out of business. Some poultrymen hung on to the old fowls for some years but finally this splendid fowl passed out.

The Brown Leghorn that we have to-day is a good fowl but not the equal of the old Brown-Breasted Leghorn that these grafters destroyed.

The old Brown Leghorn was not as long in the back and not as narrow in the breast as the Brown Leghorns—with the game blood in them—of to-day. They could be put to about one and a half pounds quicker, with proper attention, than chicks of any other breed. They were just such a fowl as you describe in many respects. There is one thing that your letter clearly indicates; that is, that you know how to handle your poultry. Now go on, save the eggs of your best layers, particularly those that are short in the back and are compactly built, and maybe you will be able to get back to the old Leghorn. You will not win with them in the shows, but you will have a fowl that will do business in eggs and broilers.

Organize an association in your county. You have as good a section as you could wish. Then get your association to affiliate with the larger ones.

# ENGLISH BREEDS.

## ORPINGTONS.

The entire Orpington family was originated by Mr. William Cook, of Kent, England. They made their first appearance in England in 1886. Ten years previous to this Mr. Cook determined to produce a fowl superior to any fowl, as he tells us, in the world. The first to appear was the Black Orpington, then in 1889 came the White Orpington. In 1894 the Buff Orpingtons made their appearance. In 1897—Queen Victoria's jubilee year—came the Diamond Jubilee Orpington. Mr. Cook presented a pen of these birds to the queen, and named them in honor of this great event. Next and last came the Spangled Orpington in 1899.

BLACK ORPINGTON HEN.

There are ten varieties. The five mentioned are bred in single and rose comb, making two distinct varieties each of the five mentioned.

Mr. Cook is called the Luther Burbank of poultry culture. Each of these varieties has been produced by an entirely different system of crossing other breeds together, and yet producing the same results except in color. They have been bred in every part of the world. Mr. Cook tells us that in order to test them thoroughly, they have been bred in Central Africa and then in the north of Russia, at Archangel, on the White Sea, and also in the extreme north of Canada. They do well in any part of the world, if given a chance.

Mr. Cook claims for these fowls that they are all alike in shape, all alike in flesh—which is superior to any other fowl in flavor; that they lay more eggs—particularly in winter—than any other breed. They are very quiet and gentle, with no disposition to fly; are larger than any clean-legged fowl. They mature quicker than any other breed; good setters and mothers; easily broken off when broody; do as well on a barn lot as on a range. They are robust in constitution, and are immune from climatic changes, and that in consequence of these good qualities they are the best fowl on earth.

I will now give an account of the way in which Mr. Cook, by different crosses, built up the ten varieties of Orpingtons. He commenced in 1876 to produce a fowl, superior as he tells us, to any fowl on earth. He had been breeding fancy fowls for years and had a perfect knowledge of the characteristics of the different breeds then known. So he selected fowls that had one or more of the strong points that he wished to combine in one breed. Black fowls were very popular in England then, and Mr. Cook, having a clear idea of the fowl that he wanted, when his

Plymouth Rocks produced a few black chickens that proved to be in line with his desires, bred the pullets on Black Minorcas and the pullets of this cross he bred on a Langshan cock. Then it took nine years breeding the result of this cross together to produce the Black Orpington. He accidentally found a few Rose Comb Langshans in a shipment just arrived from Shanghai. He secured a few of them and from this cross came the Rose Comb Black Orpington. He then commenced to breed for the whites. After experimenting for some time he produced the fowl that he was after by crossing White Leghorn cocks on Black Hamburg pullets. The pullets of this cross were nearly all white. These he crossed with White Dorking cocks and after some years got what he wanted, a White Orpington. Rose Comb Dorkings produced the Rose Comb Orpington. Before the blacks and whites were perfected, Mr. Cook commenced experimenting to produce the buffs. He had then, he tells us, the best Buff Cochins in the world. They had the best egg record of any other breed. Some of them have laid forty-one eggs in forty-one days. They were not inclined to be broody, and when broody were easily broken up. He finally mated these Cochins on Spangled Hamburgs and their chickens on Dark Dorkings. Having bred these chickens for a year or two, he brought out the Buff Orpington, crossing back on the Rose Comb Black Dorkings, and he produced the Rose Comb Buff.

The Diamond Jubilee was bred like the Buffs, except that a speckled Dorking was used. The Spangled Orpington was produced by crossing Dark Dorkings on Barred Rocks, and the result crossed on Spangled Hamburgs, Mr. Cook likes the Blacks best of all the Orpingtons.

## THE DORKING.

There are three varieties of Dorkings recognized in the American Standard of Perfection: The White, the Silver Gray, and the Colored. There are but two colors in the general make-up of these fowls—black and white. The Silver Gray is black and white distributed somewhat different from the colored variety. Of course the white is a pure, clear white, without a tinge of brass or cream. In shape, size and general characteristics they are identically the same.

These fowls are purely English and are the oldest bred in the world except some of the Chinese breeds. It is a clearly established fact that they were bred in England just after the Roman invasion. Unlike all other people, the English have never been given to the multiplication of different varieties or breeds of poultry. The Hamburgs (Dutch) have been bred there in purity for years. The Cornish Games were originated in Cornwall, England, but are classed in England and America as "Orientals." William Cook, the Burbank of poultry culture, originated the Orpingtons, which fowls have in their veins the blood of almost every known breed, and are put in the English class because there is no other place for them. But the Dorking is English and is the only utility fowl that is entirely and completely English. For hundreds of years these fowls were bred in Kent, Surrey and Sussex, England, in large numbers. They were also bred in other places in England, but not quite so extensively as in the places named above.

There were two principal markets for these fowls—Dorking and Horsham, in the south of England, and for many years a few individuals controlled the entire output of the farmers of Kent, Sussex and Surrey. These men bought up every fowl that was offered for sale and fattened and dressed them for the London market.

## The Dorking.

So far as I am able to ascertain the fowls from Sussex and Surrey were always black and white, while those bred in Kent were sometimes red in color. Again some of the Kent fowls had rose combs; this, however, was the exception. The shape and general make-up of them all was identical.

This was before the advent of the Cochin and the Brahma. When these fowls made their appearance the English people, like the Americans, went wild over the new breeds. The most extravagant and marvelous things were told about them. Unfortunately the breeders all over England used the Asiatics to improve their fowls. It was about this time—the latter part of the 40s that the name of Dorking in some way fixed itself on these fowls. In a few years the Dorking had almost passed out and the mixed breed taken its place. There were some of them, however, that had been purely bred on the large estate. In 1857-58, when the Shanghai craze had somewhat abated, several poultrymen determined to save the Dorkings from extinction. They gathered all of the old pure-bred birds that they could find and in a short while the Dorking had regained its former popularity, and has continued to hold it ever since.

The Dorking is unlike all other fowls—in shape, and in the general make-up of the breed.

In England they are thought to be superior to all other fowls in the quality of their flesh. The Standard describes them as having a broad, low-set body that is rectangular in shape as viewed from the side—a long, deep keel and short legs, which give the bird a solid, compact appearance. They have straight combs and fine toes. They differ in weight. The white variety weighs 7½ pounds cock, 6 pounds hen; the Silver Grays, 8 pounds cock; 6½ pounds hen; the colored 9 pounds cock and 7 pounds hen.

They are gentle, very energetic, remarkably vigorous and healthy. They are great layers, of pure white eggs of good size. It is claimed in England that their eggs have a richer flavor than those of any other breed. They are excellent mothers. Their chicks mature quickly and are very hardy, easily raised. The flesh of these fowls is very fine-grained, unlike in this to the Asiatics, and of superior flavor. They do not stand confinement as well as the Asiatics. They do best on a large range. They can, however, be successfully bred with treatment such as is necessary for the Leghorn family. Confinement during the breeding season and then on a larger run in the fall during the moulting season. I have always thought that this fowl would be an ideal fowl for the Southern farmer. They have been for many, many years bred in a climate similar to ours and have held their own there. They have in all these long years not only held their own, but by actual test proved their superiority to American, French, Scotch, and in fact, all other breeds that were ever bred in England as a general purpose or utility fowl. They have small bones, full breasts, white flesh. They should be, and if they were better known, would be largely bred in the South.

## THE HAMBURGS.

These fowls have been bred in England for many years. In fact, no one knows when they were first bred there, or where they were originated.

As their name indicates they were also bred in Holland. It is claimed that in the misty past, they originally came from Holland and then others claim that they were imported into Holland from England.

They were bred in England and Holland for many years under different names.

## The Hamburgs.

For many years—no one knows how many—they were bred in Yorkshire and Lancashire, England, and we are told that they were exhibited at the country fairs there under the names of Yorkshire Pheasants and Lancashire Moonies. They were, in these far-away years, bred in different colors, and were known as the Penciled Spangles and Blacks. There were several things, however, about them that differentiated them from many other breeds—their shape, their combs, their color and their ability to produce eggs. While this was going on in good old England, the Hollanders were breeding the counterpart of this fowl which they called Dutch every-day layers, and the Dutch everlasting layers, and it is positively known that the Dutch fowl was imported into England from Holland in large numbers. Later on at the great Birmingham show, there was such a close resemblance between all of the varieties, both Dutch and English that the judges agreed to give them all the general name of Hamburgs, and they have since been known by that name. For years both the English and Dutch breeders claimed to have originated them. The English still claim them, but the American Poultry Association has always put them in the Dutch class. They are a very valuable fowl. They have always been the peer of any breed as to their beauty and also as producers of eggs.

They are not very extensively bred in this country, but in England they are very popular and are bred in large numbers because of their capacity to produce eggs. There are six varieties recognized in the American Standard of Perfection: The White, the Black and Golden Penciled, the Silver Penciled, the Golden and Silver Spangled.

American breeders have greatly improved these fowls. They have by skillful breeding made them as is the almost unanimous opinion of poultrymen and all others who are in the least degree interested in poultry, the most beautiful of all the beautiful fowls in existence. This has been done so skilfully too that their value as producers of eggs has not in the least degree been impaired.

This blending of the beautiful and the useful, has made them great favorites in some parts of the country, but because of the fact that their worth in a commercial way and their great beauty are not generally known by those inter-

A Nice Pullet Strain Cock.

## The Hamburgs

ested in poultry breeding; they are not bred as extensively as they should be.

They are rather small, weighing: Cock about six and the hen about five pounds. No weight is given in the Standard for them.

Their weight, shape and general characteristics are the same in the six varieties—they differ in color only.

They are described in the Standard as being a small fowl with close-fitting plumage, rose combs, rather large for the size of the bird. They are the most symmetrical in their outlines of all breeds, and stand in this respect in marked contrast with all other standard birds. They lay rather small, pure-white eggs. There are non-setters and make up the deficiency in the size of their eggs by the number that they lay.

Some of our most skilful breeders use this breed in the North exclusively. They use them as utility birds and then sell them as ornamental birds at good-paying prices.

## THE CORNISH INDIAN FOWL.

Some time since I was asked by a lady to write up the history of this great fancy and also great utility fowl. I commenced to look into its history—it surprised me to find that there was no mention of the breed in the few books on the different breeds that I possessed. I found a description of them in U. S. Department of Agriculture Bulletin 51, where they are placed in the Oriental and Bantam class, but not one word as to their history. I read what the "Standard of Perfection" had to say of them—that they were originated in Cornwall, England. That they were produced by a cross of the black-breasted Derby Red Game with a Red Aseel Game from India, and that later Sumatra Game blood was introduced. The Standard does not state who did this work, nor does it state why it was put in the Oriental class if it was originated by an English and Oriental cross.

I next consulted "The Poultry Book," and found in it an article written by Dr. H. P. Clark, of Indiana, on Oriental Game Fowls. He states that the statement made in the Standard as to their make-up is an error, because at the time that the Sumatra cross is claimed to have been made, there were no Sumatra Games in England. After a careful reading of some other accounts of the origin of this fowl I find that they all differ, and yet all are confident that their statements are correct. It seems highly probable that they were produced by crosses of several Oriental Game breeds and then crossed with one of the larger breeds, perhaps, of the Asiatics, for it is heavier than any of the Oriental breeds, and not as heavy as the Orientals.

However, while its origin is unknown, and never will

be clearly defined, it is a handsome fowl that has been greatly improved in every particular by American breeders. It is unlike any other fowl, it is unique, it is compactly built. To those who have never handled one of them, the surprise is great when they feel their weight. They deceive their appearance. Their bones are very large and continue to grow for about three years. A cock bird at three or four years old looks as though he had borrowed a gobbler's shanks. They are a great utility bird. They are good layers and superior as a table fowl. They are healthy and vigorous. Chicks are hardy, grow rapidly, and are easily raised.

Their standard weight is, cock, 9 pounds; hens, 7 pounds. They are always in evidence at the shows where they invariably attract much attention.

COCKEREL—MATED COCK—COMPARE WITH PULLET STRAIN.

## THE POLISH FOWL.

It is to be regretted that the Polish fowl—like many others of the old breeds that have come down to us from the past—have such muddled historical records. None of them go beyond the Polish in regard to this fact. In the several varieties of the Polish fowls that we have to-day we find that, while we can not trace them directly to any of the crested or topknot fowls that we know existed in many of the ancient nations, we do know that as far back as history goes the people of these nations bred fowls that, in their general make-up resembled the Polish fowl of to-day. It is well to remember that their name does not indicate that this has reference to Poland as the place of their origin, but that they are called Polish because of the fact that they have a "poll" or lump of flesh on the top of their heads on which the topknot grows. Topknot and bearded fowls in the past were bred in China, Egypt, Russia, Turkey, Italy, Austria and in fact all over Europe. One variety, known as Padua, was bred in Italy, and another in England was called Hamburg—only because they came from Hamburg and were not in any way akin to the Hamburg fowl of to-day. The Padua was a very large fowl, built up like the old Shanghai—very tall, long legs, slim body, and were of all sorts of colors, non-setters, poor table fowls, but great layers. They have large topknots, of pure white, black and mottled. They have a very decided Oriental appearance, and are thought to have come from Malay. Those bred in all other countries were smaller, and while they differed in many respects, they were alike in this. They were non-setters, splendid layers, and only ordinary as table fowls.

Now, from this mixture came the Polish fowl that is bred in England and America to-day. American breeders in the fifties bred them up to a greater degree of perfection than could be found elsewhere, and then for some reason allowed them to go backward—latterly some of the prominent breeders took hold of them and brought them up again. The Standard recognizes two varieties—crested—and crested and bearded. In each of these there are bred several colors.

Many think that they are the most beautiful of all fowls.

The black with a white crest, the white with a black crest, and then the silver and golden spangled represents—it seems to me—the climax of the efforts of the skill of American poultrymen. Besides their beauty they are excellent layers of medium-sized, pure white eggs. They are rather small, non-setters, vigorous, energetic, and in the North healthy. They have never been bred to any great extent in the South. Those who formerly bred them here are generally of the opinion that our climate does not suit them.

## THE GAME FOWL.

Every nation, as far back as history goes, except the Japanese and Jewish people, have bred and fought game fowls, and among all the nations of the earth to-day, except, as far as I have been able to find, the Japanese, "fight chickens" still. So far as I have been able to ascertain, England and the United States, or the States that compose the Union, are the only nations that have passed laws that prohibit "cock fighting," and the people of these nations have always fought them in spite of the laws which prohibit it. The Latin race to-day in Europe and South America are the greatest cock fighters that have ever lived in the world. Thousands of dollars drift out of the South American republics to Europe and the United States to-day for pit game fowls and American breeders reap a harvest of good dollars in supplying the demand for high-class pit games.

The history of the game fowl reaches back to prehistoric times, for among the Persians and other Asiatic people this fowl can be traced by tradition to over 1,000 years before the Christian era. It has been established beyond a doubt that the game fowl of to-day has its origin in the jungle fowl of India. They came down to us through the Greeks from Persia, from the Greeks to the Romans, which latter nation distributed them all over Europe.

Game fowls, however, are found in the South Sea Islands and the islands of the Pacific, but they are evidently of different origin. The jungle fowl is a small, lightly-built, small-boned fowls, very active and quick in its actions. The South Sea and Pacific fowl is larger in size and bone, slow in action and lacking in energy.

# The Game Fowl.

Cock fighting began in England about the time of the Roman invasion and although it has been prohibited by law several times since 1191, in the reign of Henry II. it was practiced in spite of the laws passed against it and several of the kings, in whose reign it was made illegal, indulged in the practice. Even good King James I. owned game fowls, and engaged in cock fighting. The ancient Greeks and Romans used bone, brass and iron spurs on the game fowls that they fought.

No breed of fowls o nearth has received more painstaking and careful attention in their breeding than has the game fowl. This careful breeding began in Persia, where cock fighting was practiced when Alexander the Great conquered that country, and in an unbroken line continues down to the present time. The result of this careful breeding is exemplified in the game fowls of to-day.

It is manifestly impossible to give a history of the hundreds of different breeds of these fowls that have from time to time appeared in England and America. They have been bred in all colors. Some of them, like the Pyle and Duck-Wing varieties are among the most beautiful of all other breeds, and yet this has been accomplished without in any degree affecting their high standard as to gameness and usefulness as fighters in the pit. In England the Earl of Derby Games for years have stood at the head of all game fowls there, and to-day their blood courses through many of our American breeds. Prior to the Civil War the Rhett Game of South Carolina stood first among the games of the South. Hundreds of other noted varieties have since come forward, both in England and America that have proved their superiority to any of the games of former times.

Now, we have been looking at this grand bird only as a fighting machine, and not as a fancy or as a utility fowl.

There is no fowl in existence to-day that surpasses in

beauty the game fowl. I have no reference to the fowl that is to-day bred in America as a show bird, with its ridiculous long legs, its unsightly, small narrow body, its long neck, capped with a long, small head, for this fowl is useless, except as a fancy monstrosity that some people think adds to its beauty; but I am talking about the fowl known generally as the pit game—a fowl of grand appearance that denotes in its every action that it is without a peer in all poultrydom, a fowl that for ages has been bred on the Darwinian idea, "The survival of the fittest," with a pedigree that is lost in antiquity, but that has come to us through all his past generations with its chief characteristics just as they were thousands of years ago.

Now, as a utility bird. Where can you find, all things considered, their equal? Their egg product has seldom been equaled. Their energy and ability to take care of themselves on any range is not surpassed by any other fowl. Their chicks, if confined with the mother hen until their heads are feathered, can be turned out on the range, where they will grow off rapidly and are seldom, with this treatment, affected with the diseases that are common to other chickens. Their flesh is more highly flavored, and of finer texture than all other poultry. Their eggs are richer than those of other fowls. For many generations past this royal fowl has been at the head of all fowls. What other bird has the stately walk, the symmetrical form, the bold alertness, the daring, haughty look, the graceful pose and carriage? Truly, this fowl is a king among fowls!

## DUCKS.

The Southern people, somehow, have never been as much interested in raising ducks and geese as they should have been. Improved ducks and geese are as much superior to the common kinds as well-bred poultry are to the barnyard fowls, and yet for years this important farm industry has been given the go-by.

Some varieties of ducks and geese are very valuable when bred for market or for eggs. In the North, particularly around the large cities, millions of ducks and geese are annually raised. The income from the product of the duck and goose farms on Long Island, N. Y., alone runs up into the millions. Ducks and geese are easily raised and are less subject to disease than fowls. Several varieties are non-setters and lay an almost incredible number of eggs. As the Indian Runner duck has attracted so much attention, I thought that several articles on this subject would be not only entertaining, but profitable to those engaged in breeding poultry.

The ducks bred in America are divided into two classes—those that are used for commercial purposes and those that are denominated "fancy." Of the latter, there five varieties that are bred almost entirely as pet stock. Seven varieties are bred for profit.

Of the former class are the White Pekin, White Aylesbury, Colored Rouen, Black Cayuga, Colored Muscovey, White Muscovey, and the Indian Runner. Of the latter variety, the Gray Call, the White Call, Black East Indian, White-Crested and Blue Swedish, are the varieties most bred. The Call ducks are quite small and are known as the bantams of the family. The white-crested are larger, but not large enough for commercial purposes; they are, therefore, on the middle ground between those bred for profit and those bred for the show-room. The Blue Swedish are scarcely known in this country. Each of these varieties has a different nationality. The Pekin comes to us from China; the Aylesbury, from England; the Rouen, from France; the Black Cayuga is an American. As to the Muscovey, they are in a class by themselves and their origin is uncertain. When a boy, I was told that they were from Russia, but as they have been found in a wild state in South American countries for many years, if from Russia, how did they get over in South America? On the other hand, they are to-day and have been for a number of years, bred all over Europe. The Indian Runner comes from India. The Blue Swedish is from Sweden, though this is disputed.

The above-named varieties, of course, do not include the wild ducks that visit our Southland every winter, nor those that are natives of the South and remain with us all the year around. We are quite familiar with our little summer ducks. The drakes of this interesting family are the most beautiful, I think, of all ducks. The females are quite plain in plumage and yet even in plain clothing are graceful and elegant in appearance. They can be easily domesticated and become very tame, but the young ducklings are very wild at first and have to be confined when they can fly for some time until the become accustomed to their surroundings.

At the head of the duck family stands the Royal Pekin. It was imported from China in the early seventies. Like the old Shanghai, it attracted attention at once. Its great value, however, was not at once recognized; but as it became better known its sterling qualities brought it into great favor, and it at once advanced to the front ranks of the duck family, and there it stands to-day.

It is a pure white duck. In shape and carriage it is unlike any other breed. Its body is rather long and somewhat narrow, and yet its breast is plump and full. Its legs are set well back. Its body stands erect with rather a long neck, beautifully curved. It is not quick in its movements, except when frightened. They are timid, easily frightened, as are all other members of the duck family.

The Pekin is the largest duck bred in this country—the largest known in any country. They can easily be made to weigh ten pounds each. They are non-setters, lay from one hundred to one hundred and fifty eggs in a year, are not subject to disease, they mature quickly, and are, if properly cared for, ready for market in ten weeks.

After reading the above description of this duck, the question naturally comes to one: Why are they not bred in large numbers in the South? They lay winter and summer. The ducklings can be raised at any season of the year. They grow off so rapidly that they can be marketed in from ten to twelve weeks. The demand for them can not be supplied. Properly dressed, packed in ice, and shipped to any city (should the home market be supplied), they will command a better price (or fully as good) than hens.

The Aylesbury duck originated in Aylesbury, England, and is bred there in large numbers.

They are not as popular in this country as the Pekin, and in some respects are not the equal of the Pekin for market purposes. They are pure white, except that like all other white fowls, their feathers yellow under the rays of the sun in summer. They are large ducks, weighing from nine to ten pounds for the drake and seven to eight for the duck. They mature rapidly, but not as quickly as the Pekin; the ducklings being strong and vigorous are easily raised. Their feathers are of a finer texture and their flesh of somewhat better quality. Their feathers are exceedingly soft, and this is one of the Aylesbury's chief attractions in regard to beauty. It is, as I have said, considered second to the Pekin, and yet when you measure their good qualities and those of the Pekin, there is not really much difference between them. They mature rapidly; are very prolific; easily acclimated; are seemingly built upon finer lines than the Pekin. They are not affected in the least by a change in weather conditions. They are of a large size, and then, because of their soft, snow-white feathers, are the most beautiful of all white ducks.

The next in order as to popularity among breeders, is the Colored Rouen. This is a French duck and is held in high esteem by American poultrymen. The only thing that keeps this duck down to the third place is its slow maturity. They are very hardy, quiet in disposition, easily raised. For table purposes they are, by some, said to excel all of the others. They are large, weighing about nine pounds for the drake and eight for the duck. In plumage they closely resemble the wild mallard and are thought to be akin to this beautiful wild species; but if this is true, they do not inherit the wild disposition of the mallard, nor the length of its wings, for they are very quiet in disposition, and are so short-winged that they can not fly.

I have been writing about the different breeds of ducks, and now I want to say something about the history of breeding and marketing of them. Some twenty years ago enterprising breeders in New England and on Long Island, N. Y., undertook to raise ducks for market. Previously they were considered a very unprofitable fowl. There was for some reason a prejudice against them, chiefly because the only ducks on the markets of this country were those raised on ponds and water courses where they fed on fish, slugs, etc., which gave their flesh a flavor that only a few people, comparatively, relished. The men who commenced to breed them, realizing this fact, saw before them the great difficulty of overcoming this prejudice. They commenced to rear their birds on dry land, for by experiments that were carried on for some years, they demonstrated the fact that ducks could be bred to better advantage by keeping them entirely away from water, except just enough for them to drink. Well, they commenced to put this new duck product on the market. It took hard work and quite a large expenditure of money, while they were gradually, as it were, forcing their dry-land duck product on the people. After some few years it commenced to bring these breeders some remuneration. They had won the victory that they had long seen by faith in the distance. Gradually the demand for ducks increased until the demand exceeded the supply, and this state of affairs obtains today.

The large duck farms are now producing an enormous number of ducks, some of them over thirty thousand yearly, while hundreds of the smaller concerns, scattered all over the North, are in the aggregate probably furnishing as many as those who breed in large numbers.

Here again we have to ask the Southern people, why have they not been in all these years breeding ducks?

Now, let us look a little deeper into this duck problem: The Northern poultrymen have had a hard fight. Victory complete has perched upon their banners. They have opened up the way for us. Will the Southern people take advantage of this splendid opportunity to help themselvces and their State by entering quickly into this profitable business?

We can raise ducks every month in the year, for ducks lay winter and summer. They endure heat better than fowls; cold never affects them. The Northern poultrymen can only breed them profitably from February to July. With great care, that requires long experience, we can grow from the hatch a broiler in six or eight weeks that will weigh one and a quarter to one and a half pounds. You can by the same skilful management grow a full-grown duck, ready for market, in ten to twelve weeks. True, it takes a little more feed, but far less trouble. For, get a duckling over the first ten days, and if you have his rations handy, he will take care of himself—eat sand and soft food, drink water and grow day and night.

In feeding and raising ducks we swing over to the other extreme from feeding and caring for chickens. Soft food, as I have always contended, will kill young chickens; hard grain will kill young ducks. Ducks have no crops, no place where they can store up a supply of food to be taken to their mills as needed; but their food goes directly into their gizzards. There it remains only a short time, and then passes into the intestines, where it is quickly absorbed and turned into bone and flesh. Hence their rapid growth. There is a great difference of opinion as to the treatment of ducks during breeding season.

A great many people contend that much better results can be obtained by keeping ducks entirely away from water at all times. On the other hand, about as many contend that a shallow pond of water one or two feet deep is absolutely essential to their well-being. A great many of the large breeders have an artificial or natural brook running through their yards. Then again, there are a few that have their duck yards located on low, flat-lying ground, beside a swamp, where the ducks can get their fill of slugs, snails and swamp bugs. The wet and dry folks are, it appears, about evenly divided, while the swamp breeders are largely in the minority. Apart from these differences, breeders almost unanimously agree as to housing, and in a general way, feeding their flocks, from the hatch to the market. Housing ducks in the North is quite a different thing from housing them in the South. The buildings required in the North to successfully breed ducks cost a large sum of money. The laying or roosting houses, the brooder houses, incubator cellars, feeding houses and some other buildings, cost fully twice as much in the North as they do in the South. This also obtains as to poultry. Incubator-hatched ducks, it is conceded by all breeders, are the most difficult to raise, and many of the large breeders have partially dispensed with them; others have entirely put them aside and now use hens. These are used in large numbers to hatch the eggs, as they never put more than nine eggs under a hen in the early part of the year, and later on, thirteen. They seldom go beyond this number. Large hens are never used; neither are small hens. The medium-sized hens of different breeds, such as Wyandottes, Langshans, and Plymouth Rocks, are preferred. With incubators a large supply of moisture is necessary, particularly just before the hatch. The heat required is not more than that used for chickens, one hundred and two degrees for ten or fifteen days, and one hundred and three degrees until the hatch. Breeders generally keep the ducklings in the incubators for two or three days with the heat at about one hundred and four degrees—of course, carefully removing the egg-shells.

## THE INDIAN RUNNER DUCK.

While the Pekin is at present raised by the breeders and attains the marketable size more quickly than the other varieties, it now has a stronger competitor for honors in the Indian Runner, a duck not quite as large, more active, more graceful in carriage, and a beautiful fawn and brown in color. The greatest known fowl for egg production. The Leghorn of the duck family will produce more eggs yearly than a two hundred forty egg Leghorn hen. The eggs are mostly nearly white, some tinted with green, and nearly one-third larger than hens' eggs. An egg record of over three hundred eggs yearly is what is well authenticated, the eggs being deposited in the night and gathered in early morning. Besides this, I do not believe any living fowl will grow as fast for the first ten weeks as these ducks. I have them ten weeks old, fully feathered, large as old ducks apparently, and weighing four and three-fourths to five and one-half pounds, still following the White Leghorn hen who hatched them, seventeen of them, each looking twice as large as she, and they are babies more timid than a rabbit, who get scared at trifles and stampede so violently as to endanger their tender bodies by tramping over and colliding with stationary objects in their course.

They need only water to drink and bathe their heads in, but are good feeders; first, a bite of mash and then a bite of sand and a drink of water. This is repeated until they are filled up; then they are as contented as a fat hog and will sit down and grow.

They have no diseases, no roup, canker or sorehead, scaly legs or wry tails.

They do not scratch, but will pick every bug out of a garden.

They do not fly; a two-foot wire fence will keep them anywhere. They are very intelligent and learn their feeding ground, roosting pen, and attendant in a short time and make little trouble. With these characteristics you can see that they are much more desirable to raise than chickens, and if their all-round qualities were generally known, the rush to get this valuable breed of ducks would swamp the market. I am glad to tell my friends these things, and hope for their own profit they will keep their eye on the Indian Runner. W. H. SISSON.

(In *Southern Poultryman.*)

## FEEDING DUCKS.

The feeding and management of ducks is almost directly opposite to the feeding and management of fowls. Ducks are less subject to disease. They never require any medicine. A little ordinary care until they are eight or ten days old, and they will take care of themselves if you put proper food where they can get at it.

There is very little difference in the manner of feeding or in the composition of the ration among the large breeders. There is a difference as to the percentage of each of the things that compose the ration but even here the difference is slight. In one of the letters received and answered recently, the question was asked: "Should fowls and ducks be kept in the same yard?" The answer to this is: They should not be kept together. Ducks have no crops; their food passes from mouth to gizzard, as I have stated previously. They require soft food only. Fowls require hard food only. Corn-meal is not a good food for fowls—ducks thrive on it, except that it must be ground very fine. There are some things that are essential to the health and growth of ducks that would kill chickens. Whole grain of any kind, fed constantly will kill a duck; soft food of any kind, will kill a fowl if constantly fed. They should therefore be kept apart. In feeding ducks, there is a difference in the object to be attained. For market, they should be crammed with food. Those to be used for breeding purposes should be fed so as to have them grow up and develop naturally. The following is about the ration used to feed ducks for market: For the first five or six days cracker-dust or crumbs and cornmeal—fine—equal parts. Hard-boiled eggs, fifteen per cent. of total bulk of meal and crackers, fine sand five per cent. of crackers and meal, mix with water or milk. Feed four times a day. From five to twenty days wheat bran, two parts by measure, fine corn-meal one part, crushed oats fifty per cent. of this bulk, beef scraps five per cent., sand five per cent., green food ten per cent. From twenty to forty-two days old the following mixture: Wheat two parts, bran two parts, by measure, cornmeal one part, beef scraps five per cent. of this bulk, sand five per cent., green food ten per cent.; mix with water to a dry crumbly state and feed four times a day. From forty-two to seventy days old the following mixture: Cornmeal two parts by measure, wheat bran one part, beef scraps ten per cent. of this bulk, coarse sand or grit five per cent., green food ten per cent.; mix with water to a crumbly state and feed four times a day.

The above directions are taken from agricultural reports. There are slight differences in these reports. The above treatment is that generally used by large breeders.

Ducks for breeding purposes are fed as are those for market, except that less fattening food is given them by reducing the beef scrap and cornmeal one-half.

The following ration is recommended: Equal parts of cornmeal and wheat bran, five per cent. each of the bulk of meal and bran, sand, beef scraps and green food.

For laying ducks, fifty per cent. by measure cornmeal, fifteen per cent. green food (cooked vegetables, such as potatoes, turnips, carrots, etc.) twelve per cent., beef scraps and eight per cent. coarse sand or grit; mix to a crumbly state with water and feed twice a day, morning and night.

Goose and Duck Pond at Belmont Farms.

## GEESE.

Geese have never been bred in the South to any great extent since the war between the States. There are many reasons for this. For, unlike the duck, which can be bred with or without access to water, the goose finds water absolutely essential to its existence. The duck lays many times as many eggs as the goose. The duck matures more quickly and while more troublesome and expensive to get to market, gets there quicker, and thereby brings in double returns. But while the above facts are well known, there are many farmers who have swampy, low-lying land that could be used profitably by breeding geese. The market for young, fat geese is always short; the demand is always greater than the supply. They are more easily raised than all other poultry, and then, being almost self-supporting, when brought to market are almost all profit. Then when the crop of feathers is gathered, while a little behind the duck as to time, he is not far (if at all) behind him financially. This, however, only obtains where one has a pond, a running, wooded brook or swamp land. There are seven varieties of geese bred in this country, as follows: The gray Toulouse, white Embden, gray African, brown China, white China, gray wild goose and the colored Egyptian.

The gray Toulouse is, as its name indicates, a French goose, where it is bred in large numbers. They are large. The gander weighs about twenty pounds and the goose about eighteen pounds. They are very compactly built, bodies short, well rounded; short back, and full breast. They are strong and vigorous as goslings but mature slowly. The body plumage is gray on neck and body, shaded to white on under part.

White Embden geese came from Embden, in Westphalia. They have been profitably bred in this country for many years. They are pure white. They are not as prolific as some other geese, but mature more quickly. They are large, weighing about the same as the Toulouse.

The Egyptian is a very beautiful goose and is bred only for ornamental purposes.

The gray African geese are, by many breeders, considered the most profitable of all geese to breed. Thy are rated in The Standard as weighing the same as the Toulouse and Embden, but specimens frequently exceed these weights by several pounds. They mature more rapidly than other geese and can be forced up to eight or ten pounds in about ten weeks. They are also prolific, laying about forty eggs during the season. They are highly esteemed for table purposes. They are gray in color, dark on upper part of the body, shaded lighter on under part.

The white and brown China geese are identical, except in color. I bred these geese some years ago and found them to be very profitable. They are six to eight pounds lighter in weight than those above described. They are very prolific, laying from 50 to 60 eggs a year. They are very beautifully marked—neck light brown, shaded darker on body, with a deeper dark brown stripe running down the neck to the body. The under part of the body is a lighter grayish brown. The white are pure white, with not a colored feather. The beauty and graceful carriage of these geese make them very attractive.

The gray wild goose is well known and largely bred in this country. They are among the most valuable and practicable birds for good raising. They are hardy. Put out in a pasture where they can get grass, bugs and slugs, they take care of themselves. They are good layers and highly esteemed as a table fowl. They are not large, weighing about ten to twelve pounds when fully grown.

## THE MANAGEMENT OF GEESE.

Geese are more easily raised than any kind of poultry, if certain conditions are met. They require very little care. At ten days old, turn the goslings out into a suitable pasture and they can take care of themselves and even during the first ten days of their lives they require very little assistance if you confine them in an inclosure where they can get water and grass. It is deemed best by many who breed them to keep them up for a few days until they are strong enough to follow the mother goose. Others never confine them, but turn them out and let the mother care for them. This applies to the ordinary goose that we see on the farms in the South to-day. Now, when it comes to caring for the other geese, those that I have been writing about, the treatment should be somewhat different. They should be confined for the first eight or ten days and fed on cornmeal and wheat bran, two or three times a day, with grass and water always where they can get to it. It is best at this age not to let them get into the water. After they are ten days old, they can be turned on the grass plot or in the pasture, where they will find water, grass, bugs, etc. They should be provided with some sort of shelter at night, near the barn or near the dwelling, and taught to come home every night. They can be easily trained to do this by calling them and feeding them on any soft food. They will soon form the habit of coming home and will come without being called or driven up. They are now practically self-supporting, if you have a pool with trees around it or a swampy piece of land, or a wooded branch for them to run on.

By having a shelter for them, they are induced to lay about the lot and they will make nests and lay in them. When they have laid about ten or fifteen eggs they will become broody. You can put the eggs under hens, about six, or if a large hen, eight; then shut the goose up in a dark place where she can get water, but no food, and in two or three days she will be laying again. You can repeat this with the second laying, and then let the goose hatch the third hatch. In this way you can get three hatches instead of one. This does not apply to the white and brown China geese, for they do not sit. Of course you can use incubators, with about the same temperature that will hatch a chicken, one hundred and two to one hundred and three, but eight or ten days before the hatch you should go to one hundred and four. It takes thirty days to incubate a goose egg, and the eggs require more moisture than fowl eggs, particularly during the latter days of the hatch. You should buy your geese in the fall of the year, when you start to breed them. Put them out in the pasture, so that they can become perfectly familiar with their surroundings. When the grass gives out they should be fed twice a day, on fine chopped alfalfa, or clover, or for that matter, fine hay. Give them also boiled vegetables, mixed with wheat bran and cornmeal. Put in the ration about five per cent. of the bulk bran and meal, fine sand. Now, you see how small the expense and how little the trouble there is in raising geese. Each goose will give you about one pound of feathers a year. The feathers will pay you more than the cost of the feed and the geese are therefore all profit.

## CAPONS.

Latterly, I am quite sure, no subject that has been up for discussion among the poultry fraternity has been of greater importance than the brooder question. So far as using heat in a brooder is concerned, I settled that question several years ago. I found out that many chickens were killed by overheated brooders, and that chickens could be raised without heat other than that which comes from their own bodies, where you would put enough of them together, without crowding to keep themselves warm. Almost everybody disagreed with me from the first and continues to disagree with me even unto this day. Now, I have always known, and all agree with me in this, that the best brooder on earth is the hen. "But," say we all, "the supply is always limited, and when needed most the demand is always far in excess of the supply." Very true, and I fully agree with everybody in this. Now, some are for the kerosene brooder and others are for "cold storage" brooder. Well, there must be a middle ground somewhere between these two extremes to which we can all come. Well, here is the middle ground that will end old Dominick's troubles and decrease the sale of kerosene: Why not use Capons? A severe spell of illness about a year ago put me out of the chicken business. I am rapidly regaining my health. When entirely recovered I will be back into it again, and as long as I continue in it, the capon will be the only brooder that I will ever use, to the limit of their capacity. "But it is so hard to make them carry chickens," some will say. Not a bit of it, for it is easy. Take a capon of any breed except the large breeds, about a year old, or over, put him at night in a nest box, made so low that he can not stand up in it; put a little excelsior in the bottom, and give him the chickens. The nest should be placed in a coop where neither capon nor chickens can get out. Keep feed and water in the coops, and in two or three days he will generally mother the chicks. Sometimes this fails, but not often. A medium-sized capon will care for about twenty to twenty-five chickens. I have not yet patented the "capon brooder," so you are at liberty to use it if you wish to do so.

TWO FARM-RAISED PETS.

## TURKEYS.

In a short article like this, it is impossible and unnecessary perhaps, to give a description of the different varieties of turkeys that were found in America when this country was discovered. There were, however, four and perhaps five different varieties, or breeds, differing in color and shape. The variety that we breed is that which was found in this part of America, changed perhaps by domestication or crossing with the breeds farther South. The wild turkey found to-day in our swamps is just as it was originally. The turkey that is now bred in the South is much improved in color and in vigor by a cross on the wild turkey. The young are stronger and more easily raised. Young turkeys are the most helpless of all our poultry. They need careful attention. For about two weeks they should be kept in an inclosure. They should be housed at night in comfortable, dry quarters, and confined until the ground gets warm in the morning. Like puppy dogs, the young turkeys' worst enemies are internal worms; they are also subject to indigestion—but if they receive proper attention for three weeks, there is very little trouble in raising them. They should be kept in the dry ground with fine natural grit in it—a sandy place—and fed on wheat bran, having water before them in shallow pans, arranged so that if possible they shall not get wet for two or three days.

Small grain should be given, but no cornmeal or cracked corn. When about a week old, they should be given a small quantity of the common "Jerusalem oak," a weed that grows all over the South. The leaves can be given green or dried. It should be fed with wheat bran. The best way is to mix with bran and rub between the hands, dampened a little, then stuffed down their throats. This will remove the worms and also cure indigestion. After they get stronger, it should be kept up. After two weeks their range should be enlarged, when they can be fed on small grain, wheat, oats or any grain except corn. Feed every morning and be sure to feed every evening. A grass plot is a splendid place now for a small run, for they will need bugs and insects. They can now, when a month old, have a larger range, but they should, while young, never be put out on wet grass. Turkeys require a large range, and when about two or three months old, they can be turned out and allowed all the range that they can get; but be sure always to have something for their supper when they come home. Wheat bran, mixed with the "Jerusalem oak" should be fed, dampened, until they are six months old. Follow the above directions and you will make a success with the best-paying poultry that you raise.

THE PROUD TURKEY—AND RIGHTLY SO.

## THE GUINEA FOWL.

Here is a very valuable member of the domestic poultry family. I say "domestic," and yet, while it is generally so considered, it has never been completely domesticated.

This interesting fowl has been known for many years, and back in the times of Rome and Greece it was highly prized as a table fowl. It was valued among these ancient people for the delicacy of its flesh and the rich flavor of its eggs. It was bred in ancient times all over Europe, and then in the dark ages disappeared entirely. It is a native of Africa, where it is found to-day in a wild state, not exactly as we see it on our farms now, but so little changed that should we see one from its native wilds, we would at once recognize it as the ancestor of our "potrac" friend that is with us to-day. The change in plumage is slight. The markings of the head are different. The legs are of a different color from its wild cousins, but its disposition is still that of the barbarian. It takes on civilization slowly. In fact, if left to its inclination it quickly returns to its wild state. After its disappearance from Europe, it was found in the West Indies and reintroduced into Europe. It was evidently taken to the West Indies by Europeans from Africa. In many of these islands it is to-day found in large numbers as wild as it is in Africa, and is hunted as game. This is also the case in England and some other parts of Europe, where they are found in game parks of landed proprietors.

The guinea has many bad traits of character, and many that are very good. It is to be regretted that it is not bred more largely in the South than it is, for it is really a very valuable fowl to the farmer, and if properly handled can be made to pay well. It is almost an impossibility to raise chickens during the summer in the South, and as the guinea commences to lay in June, and continues until some time in August, their chicks can be bred very profitably to be used as broilers and fries in the fall. They are very easily raised if hatched out in July by using hens. The guinea is a bad setter and a careless mother, except where they are allowed to lay in thick woods and never disturbed. Their young are hard to control when this is done. By using hens and raising them about the yard, they become more civilized. When hens are used to raise them they have to be closely watched at the hatch. The hen should be confined in a close place, where the little guineas can be closely cooped for five or six days—or until they become used to the cluck of the hen. If not confined they will run out from the nest and go aimlessly on until they die from exhaustion.

The little fellows when first hatched should be fed on grass, or other small seeds, until they get strong enough to follow the hen. They should also have a small quantity of meat—green if possible. Green food is also required. They should be kept in a dry place. The guinea is the sworn enemy of the entire feathered tribe. They, if in large numbers, will keep the hawks away from chickens, and it is said that Cuffy will not steal the chickens when a guinea is roosting with them. They keep down the usual crop of insects, and eat so freely of the seeds of weeds about the farm as to almost destroy them.

## DISEASES OF POULTRY.

You will notice that as I describe the diseases and give remedies in this book that I recommend largely the use of charcoal and sulphur. Charcoal is a powerful disinfectant and deodorizer. Sulphur purifies the blood and therefore the entire system. It will keep lice and mites off the hen if fed in a large quantity. This is not best for the hen. It will put her out of condition and make her stop laying, but in small quantities each day will benefit her, particularly in the moulting season.

I have said a number of times that poultry should have before them always a box with fine charcoal at the bottom, with sulphur sprinkled over it, then wheat bran on top of the sulphur. This box should be about three or four inches deep, and a piece of poultry wire of about two to two and one-half inch mesh on the bran to keep them from scratching it out. They will eat the dry bran and get enough of the charcoal and sulphur to keep them healthy.

### ROUP.

Now, roup is one of the most troublesome diseases known to poultrymen. It is the disease that affects human beings; is called catarrh. It is confined entirely to the head in its first stages, and is caused by a cold. In its second and third stages, it affects the whole body. It can be cured in its last stage, but the fowl then is of very little value and totally unfit for breeding purposes. Its first symptoms are a slight discharge of fetid matter from the nostrils; eyes a little inflamed, and at night it is restless and makes a noise as though it were choking and gasping for breath. These symptoms after two or three days become more pronounced. The eyes become swollen, a white substance like gristle forms in them. The throat becomes sore; and a very offensive odor comes from the mouth. The disease rapidly increases in intensity, until it has spread through the entire system. In this last stage it is almost impossible to cure it. They finally become blind and die.

Now, a practical poultryman can just glance over his flock and detect this trouble in a moment, and with little effort arrest the disease and cure it. If you had piled up before you a copy of every poultry paper in the United States and would glance through them to find remedies for diseases that poultry are subject to, you would be amazed at the different ideas that folks have of their nature and at the remedies used for them. A man in Connecticut once sold roup pills at fifty cents a box all over the country. They were utterly worthless. But he had gathered in a large amount of money before he was exposed. All of the old poultrymen know his name. He is yet living, but is not in the roup pill business now. Every remedy that I recommend in these articles has been tested by me. They are based on common sense principles and are not for sale. They are gladly given to any one who asks for them. When the first symptoms of the roup appear, give the fowl five grains of quinine and with a small syringe, such as you use to fill your fountain pen, syringe each nostril and slit the top of the mouth with kerosene oil two or three times a day. Then give a pill as large as your finger and half as long of fine charcoal and sulphur, equal parts, with lard enough to make them unite. Give this once a day for two or three days. The fowl should be kept apart from the flock. This treatment will cure roup in its first stages. In the second and third stages substitute diluted chloro naphtholeum for keresone oil. One part of chloro to twenty-five or thirty parts of water; in severe cases one part of chloro to fifteen or twenty parts water.

## SOREHEAD.

I do not intend to tell you what causes sorehead, for I do not know, and I do not stand alone in this, for no one that I have heard from in the poultry business is in the least degree better off than I am. Of course, every one who has come in contact with it has his "theory" in regard to it, and most have proved, to their own satisfaction, that they know all about it; but unfortunately, they can not prove this to the satisfaction of other folks. That it can be cured is a fact; but to do this requires prompt action on its first appearance. Let it get a good hold on a flock of birds and it is almost a matter of impossibility to arrest its ravages. This I am absolutely sure of, that by keeping charcoal and sulphur before the little chicks and grown fowls always, you can prevent its appearance in the flock or mitigate its severity to such an extent, that by applying remedies given below you can banish it from the flock. It almost invariably makes its first appearance among the little chickens. Looking over them some morning, you discover a few small wart bumps on their heads. The next day the number of these little warts has greatly increased. Take one of them up and pick the little wart off and you will find that it is a cover that rests on an ugly little sore. Every chick so affected should at once be removed to the hospital. It should be given a dose, according to its size, of charcoal and sulphur, mixed with lard, and its head and mouth bathed with kerosene oil, or diluted chloro petroleum, one part of the chloro to thirty-five parts of water. If a large chicken or grown fowl, this latter can be made one of chloro to twenty parts of water. Feed on oats, dry bran, and green food until cured.

## PIP.

I have just answered an inquiry as to pip, but I will repeat it here. Pip is caused by improper feeding. The fowl has indigestion. It can be quickly cured by a change of diet and a dose or two of cooking soda. Confine the fowl in a coop and feed on oats or any grain except corn, very light; or better stop all feed for a short time. Do not follow the barbarous practice of pulling half of her tongue off and thereby causing her to suffer so much agony. This will cure her, but the remedy is worse than the disease. It only makes her unable to eat and that cures her indigestion.

Both in sorehead and pip, fowls have fever and should have quinine, five grains for a grown fowl and less according to size of the chickens.

Fowls almost invariably have fever when sick because they never sweat.

### LIVER DISEASE.

We seldom see any mention in the poultry papers or bulletins about liver troubles in poultry. Now, when we come to think of it, fowls are, in regard to diseases, much like human beings. They are often affected with indigestion; so are you and I. They have colds that run into roup. You and I have colds and it runs into catarrh. Chickens have apoplexy; so do folks. They have vertigo, and we do, too. They have pip; we do not, but then, perhaps, sometimes when we say ugly things that we should not, maybe a good case of pip would have helped us. They have sorehead. Well, you and I never had that, but we know lots of people that folks say have sorehead. Did you ever see any one with sorehead? A man or woman that had fallen out, apparently, with everything in this beautiful world, who never saw any good in anybody and who made you feel so uncomfortable that when you finally got out of reach of their tongues, silently wished that you would not soon meet with them again. Well, you can cure chicken sorehead, but it seems impossible to cure it in men and women. So as to sorehead the chicken has the advantage of the sorehead folks. But let's see about liver troubles in chickens. As to the cause of this disease, it may be malaria, as the doctors tell us when our liver troubles us, but that there are other causes I am quite sure. Improper food or too much food, we think, gives us indigestion. This is true, but I have yet to find any one who had indigestion with their livers in good, healthy condition. Liver trouble in chickens is the result of overfeeding with improper food. Keep your fowls always a little hungry, at all times with green food where they will always have access to it, and they will not be troubled with indigestion or a disordered liver. Poultrymen of experience know how to treat this disease, and also how to tell when a fowl's liver is out of order. The comb and wattles look flushed, are dark, deep purplish red. The head and eyes look just like the face of men whom you sometimes meet, and your first thought is apoplexy. They eat as much as their crops will hold, and yet in the morning their crops are half full. Their blood vessels are full, with a sluggish circulation. One or two grains of calomel, followed by a dose of oil (two teaspoonfuls) and five grains of quinine (just what the doctor would give you), will soon cure them of the liver trouble and at the same time cure their indigestion. Now, you see that you can in almost all cases of sickness among chickens apply the same remedies in small doses that would cure a man.

## TO THE BOYS ON THE FARM.

I was reared in the Sunny South. In my youth I lived about half the year on a large plantation and the other half in a city; then after the "unpleasantness" in the early sixties, on a farm for a while; then from the farm to commercial life. A boy can never drift away from his first love. If he was brought up on a farm he may get out into the commercial world, make money, live in a fine house and congratulate himself that he has made a business success amid the busy city life, but meet him at rest, commence to talk to him about business, and he will discuss business with a wearied look that denotes his inner thoughts. Now change the subject, mention some trivial incident that occurred on the farm in his younger days, and then see how quickly the wearied look disappears. His face has completely changed; at once he is—as it were, transformed. He is not now the wearied business man; he is the joyous country boy. He willingly banishes business from his mind and with a bright, glad look upon his face, will sit and talk of the bygone days, when he was as happy as happy could be on the farm. Deep down in this successful man's heart he is thinking to himself: "It would have been best for me and for my children if I had remained on the farm." Every successful or unsuccessful business man who was reared on a farm who may happen to read this will tell you that I have pictured his experience exactly in the above statement; and yet boys on the farm to-day are making every effort to get into the cities and towns, endeavoring to cast aside the freedom that they have on the farm for the slavery that they will go into in commercial life. The city life is so bright and beautiful to the farm boy, only because he does not think of the yonder; he is thinking only of the now. If he could only be induced to stop for a moment and look around about him at the old men who when as boys left the farm for the city and know that out of perhaps one hundred of those who left the farm for city life only one has made a success, while ninety-nine have made a dismal failure! On the other hand, let him look at the farmer boy who stuck to the farm. One or two have gone down, but 98 or 99 are free men—nobody to boss it over them. They have possibly small farms; some of them have large flourishing farms. They all live comfortably. They own the farm and the stock. They labor hard for five or six months in the year and then it is light work, getting ready for the next crop. Now, meet one of these sturdy farmers and ask him if he made a mistake in sticking to the farm; ask him if he would have been better off had he gone to town to make a living; watch his face. Why, he is so astonished that he can hardly reply to you. He gets up from his chair, looks you in the eye in order to see if you are not joking, and then nearly yells out: "No, siree; I would not give up my freedom on this little farm for every lot in town, if I had to quit the farm and live there."

Boys of the farm, an old man is saying to you: "Remain on the farm." If you take his advice, in after years you will rise up and thank him for this piece of kindly advice.

## QUESTIONS AND ANSWERS.

Dear Uncle Dudley:

Please advise me whether it is advisable to set eggs from hens that have been fed beef scraps. Does it interfere with the fertility of the eggs? Thanking you for your reply, I am yours very truly. J. L. R.

Colbert, Ga., March 22, 1910.

---

The above question can only be answered by going into the general treatment of laying hens. I believe that E. W. Philo is the first man that ever suggested the idea that feeding laying hens improperly was largely responsible for the death of fully matured chickens, just as they were ready to hatch, but Mr. Philo in another part of his book, tells how he liberates the chickens that had been bred according to his methods. And so it appears that he has not eliminated the trouble entirely by feeding according to his system. I am perfectly sure that you can get better results in the hatching and raising of chicks from properly fed and cared for hens. I have always been opposed to feeding food to laying hens that tended to fatten them. I have fed as little corn and meat as possible to laying hens, not more than once or twice a week; but oats, wheat and other small grain, just what they would eat up clean, all the green food they would eat, and then a bountiful supply of wheat bran. This latter should be fed in this way: Make a box about three or four inches deep of a size suitable to the number of fowls that you have; put in the bottom one inch of crushed charcoal; on this sprinkle about one-half inch of sulphur, then fill with wheat bran. A piece of poultry wire of about two-inch mesh, fitted in the box on top of the wheat bran, will prevent them from scratching the bran out. Keep this in a dry place, where they will always have access to it. Laying hens should always be kept busy; put their grain food on a soft place in the yard, spade it in and make them scratch for it once a day; or under straw chopped fine, so that it will give them some trouble to find it. Keep this idea always in your mind. You should make fowls that are confined to a small yard work as hard to get food as those out on a range do to get theirs. A little beef scraps once or twice a week will help them.

---

DEAR UNCLE DUDLEY:

In yours replying to T. L. Marchant, you say: "Corn in any shape will kill them." Our dealers sell "Purina Biddy Feed." It is principally yellow corn cracked fine, mixed with other seeds, but seems to be mainly corn. You say: "After ten days feed on commercial feed." Would you eliminate "commercial feed" if it contained any corn at all? I'm making a scrap-book of your suggestions and advice. It fills a long-felt need, especially in these days of high prices, when "eggs is eggs." V. L. S.

Waycross, Ga., April 14, 1910.

ANSWER: In all of my articles, I have only given my "experience," as the old-time Methodist used to say. I have been for many years experimenting with chickens and studying their natures and their needs, not from a "scientific" standpoint, but in a common-sense, practical way. I have found that I can keep the little fellows perfectly healthy by eliminating corn entirely for ten to fifteen days and substituting small cracked rice, or cracked wheat, or any small grain. I found out long ago by actual experiment that corn will sour when wet quicker than any other grain, and that chickens fed on it would have indigestion and die, not all of them, of course, but it is almost impossible to raise all of them. This is particularly true as to incubator-hatched chickens. Now as to commercial chicken feed, almost all of it has cracked corn in it, but some of it has very little corn. After fifteen days a small quantity of cracked corn fed with other grain, will perhaps not hurt them, if they have wheat bran always before them. I feed as little corn to large fowls as possible, except when I want to fatten them (old fowls), but it is not a good feed for laying hens.

DEAR UNCLE DUDLEY:

Two weeks ago I bought a trio of single-comb Rhode Island Reds. The hens were both laying when they reached me; all seemed to be perfectly healthy. One hen acted as though she wanted to set and soon showed signs of being sick. She was very droopy, and on examination I found that her crop had not been emptied during the night and breath was very foul. She had no appetite, but drinks quantities of water. She did not seem to get any better. Please tell me what to do for her, or the next case, and also tell me what was the matter with her.

Thanking you for the help that I have already received and for all that I know is still in store for me, I remain,

FRANK HARRISON.

ANSWER: Your hen has sour crop; cause, overfeeding on (I suspect) soft food. Fill her crop with warm water, and then hold her head down, gently pressing the crop until it is empty. Repeat it in one or two hours if not entirely emptied. Shut her up in a coop with her feet on the ground. Give her a little cooking soda and five grains of quinine. Feed her very little for a day or two and not much water.

DEAR UNCLE DUDLEY:

Your articles on the subject of poultry-raising have in-

Fourteen Weeks Friers Photographed in July.

terested me very much, and in following them I have become confident that you are a thoroughly posted man on this subject; consequently I come to you for some advice. My position is this: For four years I have been sick and have been able to do very little work of any kind during that time. I have spent a great deal of my time in studying poultry management in various magazines, books and journals, but I have had absolutely no practical experience. Now, my health is improved and I believe that light outdoor employment will help me to effect a permanent cure.

If you can advise me on the following points, I will certainly appreciate it:

1. About how many chickens can I accommodate on one-third of an acre?

2. Are not the brown or white Leghorns a very good breed for this climate?

3. When is a good time to begin incubation?

Any other information will be kindly appreciated; also notify me if there is any charge for this information.

C. W. STROZIER.

---

I am glad to receive the above letter, and take great pleasure in replying to it. In my opinion, outdoor work among the chickens will benefit you. I sincerely hope that it will.

1. This depends upon circumstances. If you were thoroughly up on handling chickens, you could manage a large number, but you say in your letter that you are not; therefore, my advice is that you get a few, say, eight or ten, white or brown Leghorns and a fifty-egg incubator and work on them through the summer, and then you will be able next fall to handle three times this number. Save all of the first hatch of pullets and they will furnish you eggs for late fall use.

2. The white or brown Leghorn do well in the South.
3. September, January and March.

Of course, there is no charge. It is a pleasure to help you. Should you get in trouble, call again.

---

Dear Uncle Dudley:

I follow up your points and clip most of them for future reference and am after a little more information than you have, up to this date, published.

I want to go into chicken-raising mostly for egg-production for market purposes, and can buy pure-bred Leghorns at about one day old. Now, are all chicks subject to what is known as the pip, or is it just occasionally that some have it? What I mean, does nature cause them (each one) to have the pip, or must I look for it on all? And what will cure it? Must it be pulled off? Also, how old would the pullets from pure-bred Leghorns have to be before they commence laying?

Now, while I intended Atlanta to be my headquarters, yet causes are shaping themselves so that I may either have to go to Jacksonville, Fla., or New Orleans, La., and I wish to know if climate in those places would be good for young chicks. Mosquitoes and sand fleas are terrible there, and I am informed (not by expert authority) that sand fleas are very injurious to chicks, causing great mortality among them by going into their nostrils up to the brain, thereby causing death. I will appreciate any information you will give and continue looking for your dope.     C. H. B.

Atlanta, Ga., April 3, 1910.

---

Answer: Pip is caused by indigestion and is quickly cured by shutting the fowl up and feeding it on oats and dry wheat bran. Give it a small lump of cooking soda about the size of a grain of corn, one dose, morning and night, which will generally produce the desired effect, but do **not** tear the poor bird's tongue to pieces. This barbarous practice will effect a cure because it stops the fowl from eating. Shutting it up for a week would produce the same result, but you can cure it in two or three days by the remedy given above. Five grains of quinine will help.

All fowls fed on soft or any other indigestible food are subject to indigestion in all its various forms.

Brown or any other of the Leghorn family are fine layers. The chickens mature rapidly. Those hatched in the early spring will frequently lay at three months old and have been known to lay a little earlier. They will be perfectly at home anywhere in the South. Chloro naphtholeum, one part to twenty-five or thirty parts of water, sprayed all over the fowls and around the premises will get rid of mites, fleas and vermin generally.

---

DEAR UNCLE DUDLEY:

I have enjoyed your letters and watch very closely your comments so as not to miss them. We have "hen fever," but a mink cooled it off considerably when he got eight of my fine Rhode Island Reds that were ready to wean. How many hens can you use with one rooster? I have read that fifteen was a good number. Very respectfully,

MRS. A. L. D.

---

ANSWER: You can run ten or twelve hens with a young rooster, if you are breeding any of the small breeds. I prefer ten even with them. With the large, heavy birds, never use over eight. I think that six would insure a large number of fertile eggs.

---

DEAR UNCLE DUDLEY:

In spite of your kind answer of a recent date to previous question, I do not want to trouble you too much; at the same time, if you do not mind telling your readers what is the best way to stop hens from being "broody," I should be obliged, and no doubt others would also. When compelled not to sit, what advantage is gained? Do they lay sooner than if permitted to bring out a setting of eggs?

IGNORAMUS.

---

ANSWER: Your apology for calling on me for help in the poultry business is accepted. It, however, need never have been made. To break a hen from sitting, shut her up in a coop and put a male with her, is the quickest way that I know to accomplish this. Moving her to some other yard where she has never been before will stop her. Many poultrymen contend that a hen will produce more eggs in a year if you will let her produce one brood of chicks than if not allowed to sit. They contend that it gives her a rest that, in accordance with natural laws, enables her to lay as many or more eggs during the year. My own opinion is that hens should be allowed to raise a brood of chicks once a year. I believe that they will continue for a greater number of years to be profitable. This, of course, does not apply to non-sitters.

---

DEAR UNCLE DUDLEY:

I have been reading with much interest your helpful pieces and am coming to you with my trouble, hoping you can help me. I have lost two fine hens that seemed to be egg-bound or to have had a broken egg in them. They are weighted down behind and walk upright like a duck, or as if their backs were broken. They stop laying all at once, but continue eating heartily. In the first place, they are too fat, having a free range. I applied several remedies, but one of them died. I cut her open and found the egg broken, but with a very brittle shell. The other hen has been this way about a week and seems to get no better or worse. Am enclosing postal for reply, knowing you can help me. O. W. BLEDSOE.

---

ANSWER: You have not told me this, but I am quite sure your hens roost too high, or have been subjected to rough handling. You say that they are too fat. Overfeeding laying hens is a source of much trouble among beginners in poultry breeding. You will have to lessen their feed or you will have more trouble with the hens, and the chickens that you hatch from their eggs. You can remove a broken egg from a hen if the removal takes place within twelve to twenty-four hours. After this time inflammation sets in, then I know of no cure for it. Better cut her head off—if you discover the trouble before twenty-four hours, anyway. After fever sets in she is not fit for food.

Lower your roosts. They should never be over three feet high. Two feet from the ground is best for large, fat hens, and three or four feet from the ground for the smaller and lighter breeds.

---

DEAR UNCLE DUDLEY:

Having read with great interest all your points on poultry, I consider you an expert in the business, and being greatly interested in poultry, I decided to ask your advice.

I am using a small sixty-egg incubator for first experience. I got only seventeen chicks from sixty eggs. I broke all that did not hatch, according to your method, and found a chicken in all but five. They were all fully developed and all seemed to have died about the same time. The shell was about two-thirds full.

I began with one hundred and two degrees and increased to one hundred and three degrees; kept all ventilators open when I could; I had a small dish of water under egg tray. The thermometer went as low as ninety degrees for probably one hour one time, and was to one hundred degrees several mornings. I aired the eggs every day at noon for twenty minutes, until the eighteenth day, and turned them twice a day until the nineteenth day.

For the past two weeks all my hens have been laying very small eggs; seldom normal size and often about one-half normal size. I have Barred Plymouth Rocks. I feed "Red Comb Scratch Feed," oats, mash feed, etc., and keep plenty of crushed bone and oyster shell before them at all times; also fresh water. Kindly give me the name of a good incubator and brooder.  E. W. HIGHTOWER.

ANSWER: I gave in an article some time ago general rules for the management of incubators. I also said that I could not, of course, give many of the little details that are absolutely essential to success. With all incubators, full instructions are furnished, and as all of them differ in construction, these instructions should be carefully followed. So, in writing that article I meant for it to apply in a general way. Your failure to hatch more than you did was evidently due to a lack of moisture in the latter part of the hatch. Try again. Read the instructions sent with the machines. On the seventeenth or eighteenth day spread a clean cloth, dampened with warm water, over the eggs, and you will have better success. You may have been feeding your hens too much and giving them too little exercise. Make them work for their living. Do not let them get too fat.

All first-class incubators give good results. Make your own brooders.

*"Heat kills more young chickens than cold."*

DEAR UNCLE DUDLEY:

I have a fine stock of chickens and the most of them are little fellows. By some means or other they have got lice on them. I would be glad if you would tell me of a good remedy that would get rid of the lice. I am interested in the poultry business and I certainly would appreciate your answer in regard to my question.

CHARLES JORDAN.

---

ANSWER: The best way to get rid of lice, mites, etc., is never to have them. Everybody who contemplates raising chicken, should very early in the spring thoroughly disinfect and otherwise cleanse chicken houses and coops. Whitewash everything but the chickens, inside and outside. Chloro naptholeum, a teaspoonful or so to the gallon in the whitewash, will help much. In reply to an inquiry made to-day, I answer your question as to getting mites, etc., off of the chicks.

---

DEAR UNCLE DUDLEY:

We set a hen on White Plymouth Rock eggs that had been shipped about twenty miles, packed very carefully in lint cotton. The hen went to sitting on the 3d and should have come off on the 24th, but on that day she hatched two, but kept sitting for a day or two after, and we decided the eggs were no good and after breaking, found the chicks ready to come out, but dead. Can you give us some reason for eggs not hatching? The hen stayed on her nest as well or better than any of our hens. C. X.

---

ANSWER: You have run up with a difficulty that poultrymen everywhere would like to solve. Some think that it is caused by a lack of moisture; others think too much moisture during the last four or five days. Others attribute it to improper feeding of the hen that laid the eggs. If the dead chickens filled the eggs there was too much moisture and if they were shrunken, too little, in the last week of the hatch. A hen set indoors, is apt to have the eggs too dry. I always springle a little water over the eggs on about the eighteenth day, but occasionally I have had about the same bad luck that you have had. See my recent article about saving the chicks on the 21st day.

---

Dear Uncle Dudley:

Could you please tell me what breed you consider as the best for raising fryers? Which breed reaches the greatest weight soonest? I have an incubator and would like to raise fryers in a small way. I have been a most reader of your comments. W. S. T.

---

Answer: Orpingtons, Rhode Island Reds or Langshans seem to be preferred as best adapted to raising early broilers, but some breeders prefer others of the large breds. Much depends on "the man behind the guns."

---

I have started in the poultry business in a small way. Have about one hundred little fellows, which I am having considerable trouble with. They will droop for a day or two and die; most of them appear to have symptoms of cholera, while others appear to have swollen crops, either full of wind or water. I put ten or twelve of the affected ones in a separate coop yesterday and they are all dead except two. All of these chickens except twenty-five were hatched from incubator about three weeks ago and seemed to be doing well until about five days ago. I had seven fine breed hatched from a hen ten days ago. All of these have died except two.

Would very much appreciate any information you can give me in regard to this trouble and treatments for the same. T. L. Marchant.

---

Answer: Your chickens are suffering from indigestion. You did not commence right with them. I rather think that you have been feeding corn in some shape, either meal or grist, or perhaps, some soft "prepared to make chickens grow" stuff. They should have been taken from the hen or incubator and placed on nice, clean sand for twenty-four hours after hatch. A little wheat bran on the sand will not harm them. After twenty-four hours, feed fine grain, cracked rice or wheat, green food and clean water. Feed them a little at a time for eight or ten days and then on the commercial feed. The life of a chicken depends entirely on its treatment for the first eight or ten days of its existence. Corn in any shape may kill them.

---

DEAR UNCLE DUDLEY:

I guess I am one of the most "crankety" chicken cranks you ever heard of, though I have had very little experience in the business. This, of course, counts for my calling to you for aid. I contemplate keeping, another season, about sixty hens, consisting of the following breeds: Single-Comb Brown Leghorns, Single-Comb White Leghorns, Single-Comb Barred Plymouth Rocks, Single-Comb Rhode Island Reds and Buff Orpingtons. Now, what I want to know is this: What height should I have my fence to keep the Leghorns enclosed? The larger birds, I think, I can manage. I want to fence off my orchard and keep my birds in it. Will build portable houses for roosting purposes; these will be large enough to accommodate ten birds. Can I breed successfully from the cockerels and pullets which I am raising now another year without new blood? These came from a standard poultry farm and hatched in March. Any other information which you can give will certainly be appreciated.

W. S. LANDRUM, SR.

---

ANSWER: My advice to you is to breed only one or two breeds. Select the fowl that you like best; build runs as you describe in your letter; put in each run a young cockerel (if Leghorns) and about fifteen pullets or hens, the latter preferred, for with this mating (a young cockerel and hens) you will be apt to get more pullets among the chickens. Have two pens of each breed, so that you can the next season change the mating, taking chickens from one pen, to prevent inbreeding. Then select a breed from the larger varieties and do as above, except that eight or ten hens or pullets are enough for one cock. The Leghorns are great layers, and so are the Rhode Island Reds and Orpingtons and Rocks. However, my advice is that you breed only one variety next year, and learn thoroughly how to handle a few, and then increase the number of fowls and varieties. To entirely control Leghorns, you would have to cover the runs or clip their wings. The seven flight feathers of one wing will keep them from flying.

---

DEAR UNCLE DUDLEY:

Would you mind telling some time in your articles on chickens at what age the Pekin and Indian Runner ducks begin laying, and if they are incubated at the same degree of heat and hatched in the same number of days as hen eggs? I am anxious to get your book when it is ready, and hope you will sell every poultry crank in Georgia one. It makes no difference how long you have been raising chickens, you can be taught something new. Hoping you won't be offended at me for not signing my name, I am, as ever.

---

ANSWER: The Pekin and Indian Runner ducks are non-sitters. They both mature quite early and commence to lay at six to eight months old. Some have reported the Indian Runner to have laid even before this age. The Indian Runner commences to lay somewhat sooner than the Pekin and lays more eggs than the Pekin. The Pekin is much larger than the Indian Runner. The degree of heat is the same as that used for chicken eggs, one hundred and two to one hundred and three degrees though with duck eggs it should be raised to about one hundred and four toward the last of the hatch. Duck eggs require more moisture than fowl eggs.

Dear Uncle Dudley:

Will turnip greens stop hens from laying? I have been told that they will.   V. C.

Answer: I have been asked this question before by another poultry breeder. I have lived on a farm. I always raised turnips in the fall. I have run hen on the turnip patch in the winter. I have seen them eat all that they wanted of the tops, it being the only green food that they could then get. If it ever stopped them from laying I did not know it.

Dear Uncle Dudley:

Another question or two.

Without using trap nests how can one best tell the chickens which are not laying and which are eating their heads off and should be eaten? Is it all right to kill chickens which are broody? Please give us one of your wise dissertations on the slaughter of these innocents.

I have a peculiar rooster. He stands up erect as a penguin; indeed, as time goes by, he almost falls backward, so ultra-military is his attitude. But he is not at all soldierly in his gait, for he progresses by throwing his legs out sideways in an exceedingly comical fashion. I keep him as a curiosity, and my man wants to enter him in a dime museum, out of which he thinks a nice little income could be obtained, especially if a handsome hen could be induced to flirt with him and make him show his paces.

What kind of a dislocation or other anomaly has this animal, do you suppose?

With thanks for all your wisdom spread so liberally before the public. IGNORAMUS STILL.

---

ANSWER: It takes some experience to pick out the hens that are laying in the flock. This may help you: The laying hens have red combs. They are restless. They sing sweetly, and to a genuine lover of poultry, sweeter than the folks in the grand opera. Those not laying have a listless way of conducting themselves, and to all outward appearances have "no music in their souls." Just keep watching them and you will soon be able to tell which are laying and which are not. Do not, however, be too quick to act, for some hens are on the eve of laying before their combs look bright.

Your rooster has probably rheumatism, or it may be that he has had a blow on his back. Give him a laxative, a teaspoonful of kerosene oil and five grains of quinine, and if this does not cure him, turn him over to "Jedge Briles." He may have met up with a tiger that had his eyes open.

---

DEAR UNCLE DUDLEY:

Why is it that poultrymen ship eggs for setting that are not uniform in size, color and form? I have recently handled five sittings in each of which were eggs I would not select for setting myself or shipping to others. Should eggs with ridges, bumps or with watery or cloud-colored shells be set? Can you give us some rules for selecting eggs for setting? M. T. E.

---

ANSWER: As to the size and color of the eggs that breeders ship out for hatching purposes, I have this to say: Almost all of the new breeds have been built up by the introduction of Asiatic blood, and some of them lay white and others brown eggs. Now, if the eggs are perfectly formed, the color and size amount to very little unless they are unusually small. Ridges and bumps on eggs render them unfit for incubating purposes.

DEAR UNCLE DUDLEY:

I have been very much interested in your articles. I would like to ask a question. I live in the country, and am troubled a great deal with hawks, and have been advised to feed nux vomica to the chickens to kill the hawks. Do you advise it, and would it injure the chickens?

ANSWER: Nux vomica will not hurt chickens. It is a fine tonic for them. A piece of sheet tin twelve by twelve or larger, hung on a pole so that it will revolve will sometimes frighten the hawks away. Try it, and then report as to your success with it.

DEAR UNCLE DUDLEY:

I had seven chicks hatched under a hen three weeks ago; six of them are still living and doing well. I also had thirty-nine hatched in an incubator and all are dead except six. In other words, at three weeks old, six out of seven the hen hatched are living and six out of thirty-nine the incubator hatched are living. I kept the temperature at one hundred and two and a half for two weeks; the third week the temperature ran up to one hundred and three to one hundred and three and one-half. The eggs always felt warmer than the eggs under the hen. The incubator was run as per directions. Are chickens hatched in May any harder to raise than those hatched at any other time? Some people say those hatched in May will all die. I never gave these chicks anything but grit for thirty-six hours, and then I fed them on rice, wheat bran and other

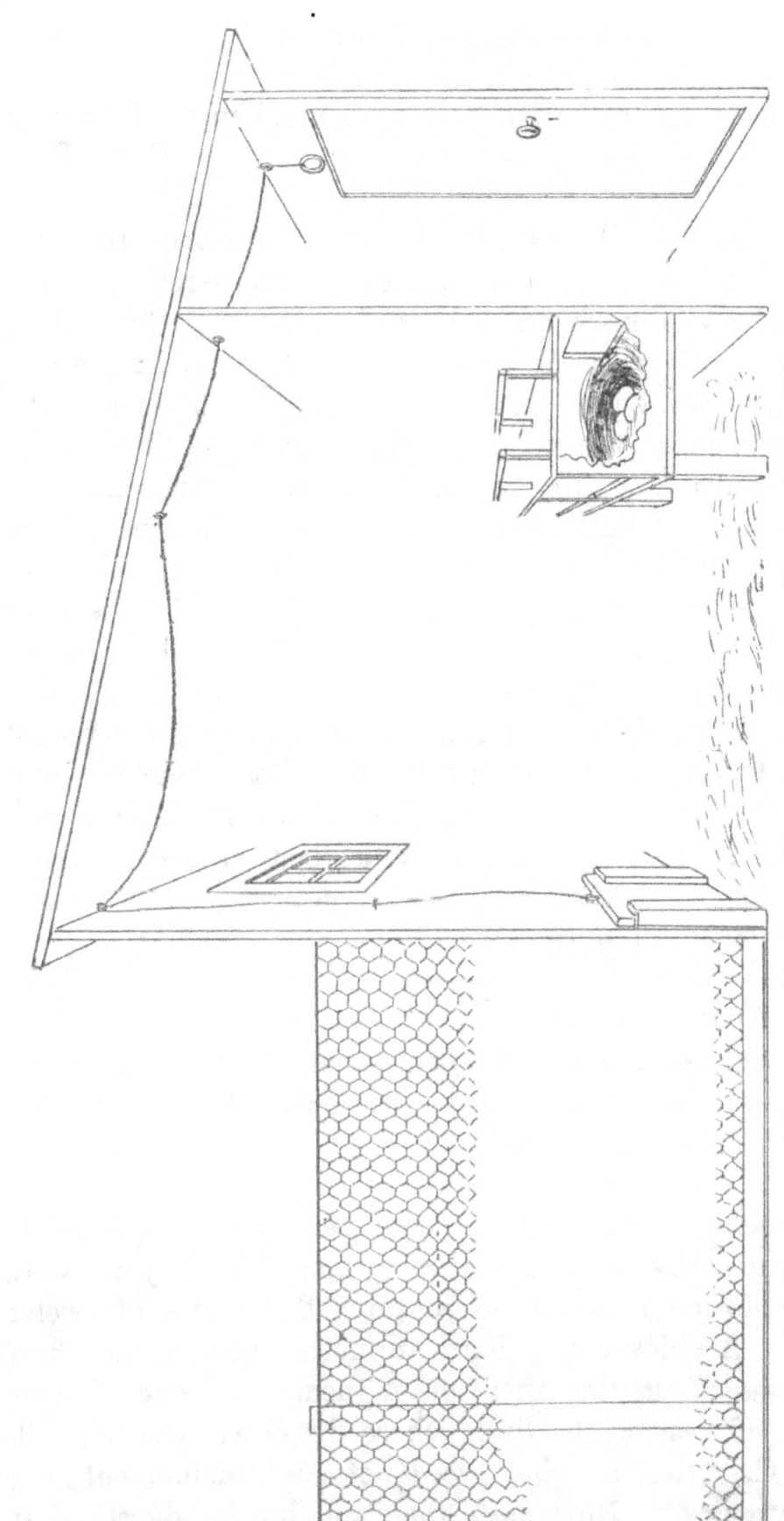

An Excellent Plan for Poultry House for Either One or More Sections.

small grain. I read all of your pieces and enjoy them very much. R. C. F.

Answer: Chickens hatched in incubators are more difficult to raise than those hatched and mothered by a hen, particularly when handled by a beginner. It takes long experience to raise chickens successfully. Do not be discouraged; a failure now, if carefully considered, will teach you how to avoid the mistakes that you have made, and, by avoiding them, bring you ultimate success. So far as May chickens are concerned, I have for years, been unable to make any great success with them. I have always found that they are more subject to disease than those hatched earlier in the year. When sorehead attacks the flock, with me, it has always commenced with late-hatched chickens. I have a good many times pulled them through all right, but it takes close watching and much trouble. I think it best for a beginner to quit getting off chickens after April and then commence again in September. Keep the hens laying, for you will make more money selling eggs than raising chickens.

Dear Uncle Dudley:
Is enclosed formula a good general chicken powder? If not, will you please give me one that will keep chickens well and also make hens lay? M. R.

Answer: For many reasons I am utterly opposed to doctoring chickens or any other animal when they are well.

The formula mentioned contains the names of twelve drugs. A chicken is a hard animal to poison, but there are some things that will poison them, and one of them is salt. I read in the Bible where the Great Teacher tells us: "They that be whole need not a physician, but they that are sick." Now, as to the stuff that is advertised in most—and I suspect all—of the poultry papers to make hens lay is, in my opinion, absolute poison, and in this way some of them will make hens lay, perhaps, but it acts upon them like morphine upon the human body, "and the last stage of that man was worse than the first."

I have never found any trouble in bringing them up to their individual capacity in the production of eggs by feeding oats, wheat, rye and other grain. I do not feed corn at all, except to fatten grown fowls for market. This treatment is, after years of experiment, the best way that I have found to "keep chickens well and also to make hens lay." Cut out all drugs. They are generally advertised to sell, and the fellow that uses them is almost always badly sold himself.

## LET BEGINNERS GO SLOW.

Dear Uncle Dudley:

Being a subscriber of the Cultivator and counting it a great help to me, I want some advice on the poultry business.

I want to begin a poultry business and want to raise chickens and eggs for the market and make profit on them. I can raise my own feed, such as wheat, corn, kaffir corn, sunflower, and expect to buy what I can't raise. Find enclosed maps of my lots which I planned 20 yards wide, 35 yards long, and build a chicken house in each one 10x10 feet, and put twelve or fifteen hens and a rooster in each one. Think I will try the Rhode Island Reds for market chickens. Will they pay as layers? And want to learn to use incubators all together. Which kind is a good one? I want a 200 or 300 egg incubator, and also want to know where to buy the best poultry book you know of for a beginner.

Do you think I can make it pay? I am willing to follow instructions closely, and my feed bill won't be so much as I can make most of it and make enough other farm stuff to buy all I need in this line. C. L. P.

Sumner, Ga.

Comment: This letter reads just like many that I have received from those who have never bred chickens according to up-to-date methods, and yet want to go into the business because it looks so easy. In going into any business you should—to make a success—have a thorough knowledge of the business.

The poultry business is just like the drug business or

the grocery or dry goods business, and should I ask you if you could just read a book and learn by it how to operate a drug store you would tell me that you could not. Well, you can not learn from books entirely how to manage as large poultry plant as you describe. My advice to you is not to attempt it for you would probably lose all the money that you put in it.

Very few people ever make a success with the first hatch with a 50-egg machine. They soon learn. You could not make a success at first with a large machine. Learn how to breed a few chickens; make a study of the business with them and gradually increase the number. Your plan for the yard and houses, so far as they go, look all right, but there are many things that you do not mention. Things that can not be clearly put on paper without a cut.

The Rhode Island Reds are a great fowl; as good as any breed for eggs and chickens. The Leghorns are great layers and with the use of incubators are profitable for eggs and broilers, to those who know how to manage the business. Any of the first-class incubators, if you follow the instructions closely, will give good results. You have a great advantage in being able to raise your feed. Fifteen hens to one cock will do for Leghorns but with Rhode Island Reds you should never have more than ten hens to each cock. You would get more fertile eggs if you used only eight hens with one cock.

Now I have written plainly what I have learned from experience. Do not "Bite off more than you can chaw." Begin with a few, learn the business and then you will make a success.

## SCRUBS AND PURE-BREDS.

DEAR UNCLE DUDLEY:

I have been raising chickens 31 years and I learn more

about them every year. My idea is to have a place to keep the little ones where they won't be trodden on by the large ones. I have always fed them on bread, chopped corn, wheat and oats, and don't feed on raw dough at all; and I always have oats sowed for them to graze on in the winter and spring of the year, and most always have grain sowed near the house for them to pick up the scattered grain that is on the ground. It don't take much other feed for them at that time. I look after them closely. If there is one to die I bury it; and don't allow anything to be thrown out for them that they ought not to have. I have a small yard to keep them in till they get large enough to run around with the hens. I had somewhere close to three hundred hatched off this year, and I have lost but very few of them. I took off eighty in May that are doing nicely. I have coops for them till they get about partridge size; then I train them to go to the fowl house, and I keep it as clean as I can. I whitewash with lime, and sometime sprinkle with kerosene oil; that kills the mites out for me and when I take a hen off, if she has any lice on her, I grease her with lard and salt, and the little ones get it when hovering. I sell lots of eggs and chickens also. My chickens are mixed, but I have very good luck with them. I have been thinking for some time that I would get pure blooded stock, but the prices of the eggs are so high by the time I raise them I could never get half what I paid for them. I would be glad to get 50 cents a dozen. I think I could make more money at that. You can't get enough to pay to raise them, much less pay $1.50 for 15 eggs. M. P.

Natasulga, Ala.

COMMENT: I am glad to get this letter. I judge that you live on a farm. You have, I suppose, lived on the farm for years. Now I am going to tell you that in my opinion you have done remarkably well with your chickens. You have, however, been breeding them in the old-fashioned way, using old-fashioned methods and have gotten the best results that possibly could have been gotten out of them by the old-time methods. But these are progressive times, and had you used up-to-date methods you would possibly with the same, or less work, doubled your output in eggs and chickens. If you will count your hens and then the eggs that they produced you would see that they do not, perhaps, give you more than seventy or eighty eggs a year each. Now, almost any of the pure breed fowls, with the same careful attention that you give your mixed-up barn-yard fowls, would have given you from one hundred and twenty-five to one hundred and fifty eggs a year.

You can purchase from some reliable breeder fifteen setting eggs for 10 cents each, one dollar and fifty cents for fifteen. Say that you raise six of them, which is a very low average. You have six fowls that are worth at least two dollars each. But put it at $1.00 each, this gives you at the very least calculation a profit of three dollars and fifty cents. Then in the up-to-date management of your flock you can get better results in the fry output. It takes you four months to produce a fry. In the quick growing utility breeds you can market your frys in ten weeks and have better frys than the old common fowl can produce.

It is very dangerous in the South, to raise chickens in the summer. You should commence in September and quit in April. In this way you would get rid of all the summer diseases. Would not have to fight lice and mites like you have to fight them to save the lives of the small chickens. Subscribe to the poultry papers, and make a start with the pure breed that you prefer and you will, I am quite sure, double the profits on your poultry.

Dear Uncle Dudley:

I am writing you on the poultry question for some advice. I have read a great deal about raising poultry in the Cultivator and journals and I find to have a certain place for them or a yard is best. I have wired yards to put my little ones in. I feed them on corn chops and wheat, and give them skimmed sweet milk to drink. I find this a good food for the little ones. I now have 225 little ones and have only lost five in this number. I have a large number nearly frying size, and I now have 128 eggs setting. I have good luck so far. I am expecting to raise about three or four hundred this year. I have read a great deal about the incubators, but I prefer setting the hen. I set two or three at the time, so as to have a larger drove, and I have to feed them well to keep them up. I have a good variety of chickens, but I want a full bred stock of White Leghorns. I have white Leghorns mixed and and Wyandottes also. What kind of chickens do you think best? I have only about forty hens.

M. P.

Notasulga, Ala.

---

Comment: I have answered many letters like the above, and whenever I read such letters, the thought invariably comes to my mind, "What a pity that this writer breeds common fowls." Anyone who reads this letter can see that the writer is capable of managing a flock of four or five hundred pure-bred fowls that would pay two or perhaps three times better profit than the mixed breeds and yet there are hundreds of people scattered all over the South who have not yet learned or "caught on" to the fact that pure bred poultry pays best.

You have had good success with your chickens so far. You did not go far enough into the details of how you treated the little fellows from the hatch to the fry size. I wish that you had. You tell us that you have fed on corn and wheat. This is all right for the fries, but I could never make a success with any corn at all for the little fellows. I never feed any corn to them until they are over twenty days old and then very little until you wish to fatten them as fries. The hen is the best incubator in the world; no man has ever made one that equals it. Setting a number of hens at one time and then giving one of the hens a large number of the chickens and having the other hens soon producing eggs is a good business proposition. No man ever made a brooder that can equal the hen as a brooder of chickens.

Now as to the White Leghorns and the Wyandottes. These two breeds were bred for distinctly different purposes. The Leghorn has been for many years, bred to fill the bill as an egg machine, and they are that and only that. The Wyandotte on the other hand, has been bred as a general purpose or utility fowl. They do not produce as many eggs as the Leghorn, but they are better as a producer of broilers and frys. Then they are quiet in disposition and make good, careful mothers. They are good winter layers and excellent for the table. My advice to you, therefore, is to select the breed that suits your business best. If you see that you can make more money on eggs alone breed the White Leghorn. If you are going to breed for frys, take the Wyandottes, but in no case mix them up, for when you cross them you have a fowl that will quickly degenerate and go back to the old "Dunghill" that will not pay much more than for the food that they eat.

Any careful reader of your letter will quickly discern the fact that with pure bred poultry you can make a large success. Write to us again, and tell us more of your method of feeding the little chickens.

## MARKET CHICKENS.

Dear Uncle Dudley:

In order to successfully raise chickens and eggs for the market, what breed or breeds would you consider best for Middle Georgia? What is the value of separator skimmed milk per 100 lbs. when used to feed chickens for the market? What other feeds should be combined with skim milk to produce eggs, and what to produce quick maturity in the young chickens? Is there a better price paid for capons than for ordinary chickens in Georgia? How many hens would you consider it wise to keep on one range, with plenty of green stuff for them to pick?

J. H. Hooks.

Warthen, Ga.

---

Answer: All poultrymen differ as to the best breeds to use. Each claims that the breed that he is using is the best, and all who are using intelligently, up-to-date methods are making money on chickens and eggs.

Many of them use two breeds. The one for eggs and the other for broilers and frys. The White Leghorns are the most popular for eggs; they hold the record as egg producers. Either of the following breeds will give good results: Wyandottes, Rhode Island Reds, Orpingtons or Plymouth Rocks. There is this, though, that you must consider. Some strains of each of these breeds seem to be superior to others of the same breed in the number of eggs produced. Some breeders claim that for some years back they have been using trap nests and have, year after year, been using as breeders only those hens that have the highest egg records in their flocks. There is some truth in this, therefore you should buy from reliable poultrymen only, and be sure that you buy the best that you can find.

I can not now give you the relative value of milk as a food for poultry. I had a bulletin giving the result of a series of experiments that were made at a Northern experimental station, but it has disappeared from my file.

In some parts of the eastern states capons bring a higher price than ordinary fowls, but there is no demand for them in the South above the market price of ordinary poultry. They are highly esteemed in France, where large numbers of capons are marketed at fancy prices. They are superior to other poultry for the table and if more of them were put on the market, they would soon be a large demand for them at paying prices. With the Leghorns you can use about ten or twelve hens to one cock, but with other breeds named not more than eight should be used.

---

Dear Uncle Dudley:

Again I'm coming to you in behalf of my chickens. Before I think we misunderstood each other.

My pullets and hens from a year old and up are affected with a very peculiar disease.

Sometimes the victim is very light, as though starving, again it is in most excellent shape.

The bowels do not seem deranged. They grow weak legged, lose sight of the eyes, combs turn dark.

Sometimes they linger for a week or more, and again two or three days does the work. Keep fresh water before them all the time. Feed chops (sparingly) oats and wheat bran. Occasionally I gave salts in water and once a week sulphur, that is if the weather permits.

Please tell me, if you can, what to do. Now, I have kept them up on light rations, giving a liver medicine but with no avail. My neighbors are losing theirs in the same way. An early answer from you will certainly be greatly appreciated.
Mrs. F. H. Scott.

ANSWER: Your fowls have evidently been poisoned by eating maggots or decayed flesh. This trouble is very difficult to cure. I have cured it in its first stages, by giving a strong laxative followed by a complete cleansing of the entire system, a good dose of castor oil is good. Some one told me recently that a dose of Epsom Salts was good. I have always been opposed to salts as a medicine for fowls, but recently I have seen it recommended by many, and where other remedies fail I would use it. Now, after the laxative has acted I would give a good dose of dandelion, root or powders. This is the best liver tonic for a fowl that I have ever used. Sulphur (in dry weather), charcoal, and lard should be given for about two or three days. These are remedies for this trouble in its first stage. When the fowl gets real sick, apply the hatchet and burn the carcass.

Clean up the entire premises. If you have done this go over the entire place again, disinfect completely—use lime freely and I think that you will find the cause of your trouble. Remove it and your trouble will be at an end.

## INDIGESTION IN CHICKENS.

DEAR UNCLE DUDLEY:

Seeing your many answers in the SOUTHERN CULTIVATOR on poultry, I write you for a little information. Last year I had over one hundred chickens hatched, and when they were several weeks old they would stand in the yard and nod as if they were trying to sleep, and would be drowsy and would not eat, and as soon as I brought them in and gave them any medicine they died in a little while, and finally I lost nearly every one. I disinfected the yard every day; used some in water, made a pen with charcoal, sulphur and bran; fed them little chick feed with Bigler's poultry powders, and in vain. I lost them.

and feel almost too discouraged to attempt again this year, but I have already set two hens, and am anxious to hear from you to know if you can tell me what the trouble was and a remedy. The chickens' bowels seemed all right, but their craws were full when they commenced to look droopy. Think I have given you full information.

Mrs. C. E. B.

COMMENT: Your chickens have indigestion. I rather think that you have been feeding them on cornmeal. It is a very difficult matter to cure little chickens when sick that have been improperly fed for the first ten or fifteen days. Corn in almost any shape is almost like poison to little chickens. They should be fed on small grain, such as wheat, rye or cracked rice for at least fifteen days. They should not be fed at all until they are about thirty-six hours old. They should be put where they can get grit, then wheat bran and fine alfalfa. Never feed them on soft food; keep them comfortable and dry; should they look a little droopy give them a little kerosene oil in the water that they drink.     YOUR UNCLE DUDLEY.

DEAR UNCLE DUDLEY:

Please explain the cause of the following:

I had a hen to die this morning very suddenly; on going to roose last night she seemed perfectly well, but this morning when I approached her she was sitting on the roost gapping as if choking, or for lack of breath, and died in a few minutes. She seemed to be in fine condition.

H. L. P.

Lakeland, Fla.

COMMENT: I wish that you had described more fully than you have done, the symptoms of this hen. You have told me only enough to enable me to guess at what killed her. If there was a large discharge from her, of a greenish, yellowish color she had cholera. If she was over fat and died suddenly without any such discharge as described above, she had apoplexy. If the former was the case, you had best disinfect and clean up your entire premises, for if you do not quickly remove the cause, you will have trouble with your entire flock. If the latter is the case, and you are overfeeding your fowls, stop their feed for a day or two. Do not give them any corn, but feed them on just a little oats and green food until you reduce the fat.

## CHICKS DYING.

Dear Uncle Dudley:

My chickens are infected with a disease that is not familiar to me. Symptoms: First, inability to pick up food, except bread, bran, and soft feeds. Presence of mucous in mouth, weakening of legs, finally a gradual blackening of the comb. Sometimes the bowels are very loose, actions dark green, touched with white. What is it and what must I do?

I keep premises disinfected, fresh water before them all the time. Feed chops, bran, oats, little cottonseed meal. Give gravel, ground shells, charcoal. In water occasionally turpentine, copperas, or carbolic acid.

Mrs. F. H. Scott.

Itta Bena, Miss.

Comment: You are feeding and doctoring your chickens to death. Well fowls and little chickens never need medicine of any kind. It costs more to raise little chickens in the heat of summer than they are worth, in the South. The proper thing to do is to stop hatching chickens in April. Commence in September and raise them all through the winter. If everybody would do this we would increase the number of winter laying hens. Cottonseed meal and corn in summer are poison to little chicks. It should be fed carefully to hens in summer.

Never under any circumstances feed soft food to little chickens until they are over 20 days old. After 36 hours from the hatch feed them only on small grain; as soon as they are dry, from the hatch, put them in a comfortable (not hot nor cold) but cool brooder. Have clean sand sprinkled on top with a little wheat bran and alfalfa meal. For the first 15 days feed on cracked rice, cracked wheat, millet seed or other small grain. Feed them on oats, wheat and other grain, but very little corn. Of course they will need sand, fresh water, fresh air, cleanest houses and constant attention.

## IN REPLY TO ABOUT EIGHT LETTERS OF INQUIRY.

I receive some times a number of letters in which the writers ask questions along the same lines, and it is frequently the case that these questions have been previously answered in the articles written or in answer to the identical questions asked by some one else previously. Then, again, some do not get a clear idea of what I really did say in some of my articles, and several times since in reply to questions. I have said that what I had written and what I would in future write was entirely taken from my experience as a breeder of chickens for a long time—something like fifty years. Some have written me nice, kindly letters, saying, "I differ with you entirely on this subject." My invariable reply has been, "I am glad that you do, for if we never have folks to differ with us we will never arrive at the truth."

Now, here are some of the questions that have been asked me very recently: "Do I understand you to say that salt will poison a chicken?" This is in answer to two inquiries.

What I have said is this: That salt in large doses will kill chickens, and even in small doses is injurious to them. Now, in the past two weeks this question has come to me from several: "Will corn poison fowls?" Now, I am quite sure that I never said that, and you who have asked that question can not possibly find it in any of my articles. "What did you say, then?" Well, I said, and still say, that corn in any shape fed to newly hatched chickens, in my opinion, was like poison to them; that they should be fed on small grain until they are at least ten days old; that I feed them on small grain until they are fifteen days old, and that cracked rice gave me better results than anything that I have ever used for little chicks.

After they are fifteen to twenty days old the commercial chicken feed could be fed, provided it did not have too much corn in it. Now, as to feeding laying hens, the commercial feed with corn in it will not hurt them, if there is a superabundance of other grain to balance it. Corn alone fed to them will quickly make them so fat that they can not lay. I have replied to several letters on this point.

For fattening poultry for market corn will give better results than any feed that I have ever used.

---

A subscriber asks:

"Why do hens eat eggs? Can you give me a cure for the habit? I would rather feed sixty-cent corn than twenty-cent eggs."

Egg-eating is a vice usually started by one or two hens. If the ringleader is caught and killed, there is usually no further trouble. Another remedy is to cut the beaks off bluntly to the quick. If a point is left, the work is of no use. Or if plenty of eggshells are to be had, feed them by the bushel (practical only when egg shells can be procured from bakeries) until the birds are sick of the sight of them.

It is probably a craving for more lime. If the house is not well supplied with nests and eggs are laid on the floor, the hen first investigates. If the egg breaks she and her mates eat it, and in a short time learn how to break and get the tempting morsel. Provide plenty of nests, plenty of lime, and watch for the chief offender.—*Exchange.*

---

COMMENT: The above was clipped from a poultry journal. I neglected to mark the name of the paper on the clipping. Egg-eating hens are troublesome customers in a poultry yard. They soon teach every hen in the yard to form the habit. It is sometimes hard to locate the ringleader until several have been taught that eggs are a great delicacy. The habit is caused in several ways, and almost in every instance by careless handling. Among the causes that have been observed by me, and that breeders have told me about are these: Putting the shells of eggs that have recently been used, with possibly some of the egg left on them where hens can get them. Soft-shell eggs that are nearly always laid at night. Feeding fresh meat freely and then stopping suddenly. Then, on the other hand, confining them in a small yard and giving them no fresh meat, or scrap meat. It sometimes also happens that they are forced to lay in an empty box, or in a barrel, where they have to fly down to reach the nest and thereby break the egg, and thus get a taste of the broken egg. They soon learn how to get into a perfect one. An insufficient supply of lime will cause them to eat eggs. I have my serious doubts about making a hen sick of eggshells by feeding them in large quantities. In my opinion they would eat what they wanted and then come back when they needed more. I know a breeder, who in order to make his hens lay and his chicks grow off quickly, fed a large quantity of fresh meat to them every day. The butcher failed to come for one or two days and almost every hen that he had went to eating eggs. He said that they got so ravenous for eggs that they tried to eat glass nest-eggs, and that one large Orpington hen actually succeeded in smashing one of these, but discovered her error before she ate it.

## SOME QUESTIONS THAT UNCLE DUDLEY WANTS ANSWERED.

First. We are drifting into making a large percentage of our hens non-sitters? It seems to be a well known fact among poultry men that you can take a bunch of pullets of any breed and from the start never let them sit and they will finally show very little inclination to take the nest. Do the next generation in the same way and the inclination to sit is lessened. Keep on this line and you will ultimately have a breed of non-sitters. Now, I have never tried this, but I have seen it so stated in the poultry books. On the other hand, I do know a breeder, right here in Georgia, who has for several years been breeding single-comb Brown Leghorns. They are of as good strain as is bred anywhere. They are not in runs, but are all in one flock, and have the freedom of the farm. He commenced some years ago to encourage them to sit, and now, as he told me, they sit and mother chickens like any other breed. He sells utility birds, chickens, broilers and fries and eggs. On many of the large poultry farms, where incubators are used exclusively, very few, perhaps none, of the hens are allowed to sit, therefore decreasing the inclination to take the nest. My attention has been called to this subject and I have been asked "why is it that my Plymouth Rock hens will not take the nest when others around me have hens that are all wanting to sit?" These complaints are more frequent than they formerly were. I would like to have some of the older breeders let us hear from them on this subject.

YOUR UNCLE DUDLEY.

www.ingramcontent.com/pod-product-compliance
Lightning Source LLC
Chambersburg PA
CBHW082330220526
45470CB00008B/2454